The Cruise of the Teddy

To my wife and shipmate

JULIE

The Cruise of the Teddy

ERLING TAMBS

First published 1933

This edition published 2020 by
Lodestar Books
71 Boveney Road, London, SE23 3NL, United Kingdom

www.lodestarbooks.com

A CIP catalogue record for this book
is available from the British Library

ISBN 978-1-907206-49-8

Typeset by Lodestar Books in Adobe Jenson Pro

Printed in Wales by Gomer Press Ltd

All papers used by Lodestar Books
are sourced responsibly

CONTENTS

ILLUSTRATIONS

Publisher's Note

We would like to thank Barbara White of the yacht *Zoonie* who met with Erling Tambs's descendants, drew this book to our attention, and provided an initial scan and edit of the text; Allan Grey for scanning and enhancing the photographs; and the Tambs family, and Anne Nygren of the book's Norwegian publisher Flyt Forlag, for permission to produce this new English language edition.

PREFACE

In presenting this unpretentious narrative of our cruise to the English-speaking Public, I hasten to avert the wrath of stern critics by stating that I do not pursue any literary aim. If my English expression seems awkward I offer my humble apology.

However, to us of the *Teddy* this cruise, undertaken in an old pilot cutter with an ever-increasing family for a sole crew, has seemed to be so full of adventurous and interesting incidents that I trust my readers will derive some enjoyment from the account thereof, even though the account be given in a language with which I am not thoroughly familiar.

I owe it to the kind collaboration of Miss J. Macleod of Dunedin and Mr. H. T. Gibson of Auckland that the worst errors of language have been eliminated without disturbing the narrative.

From a personal point of view my book has one object, namely, to provide the financial bedding on which to build a new *Teddy*, another floating kingdom in which I may continue roaming about amongst sunny isles and hospitable shores. For your little share in assisting me to achieve this aim, please, Reader, accept my sincere thanks.

I also owe thanks to numerous friends we have made during the cruise for their assistance and encouragement, and in this respect I feel particularly indebted to the kindly people of New Zealand.

Erling Tambs
Otuarumia
Hawkes Bay
July, 1932

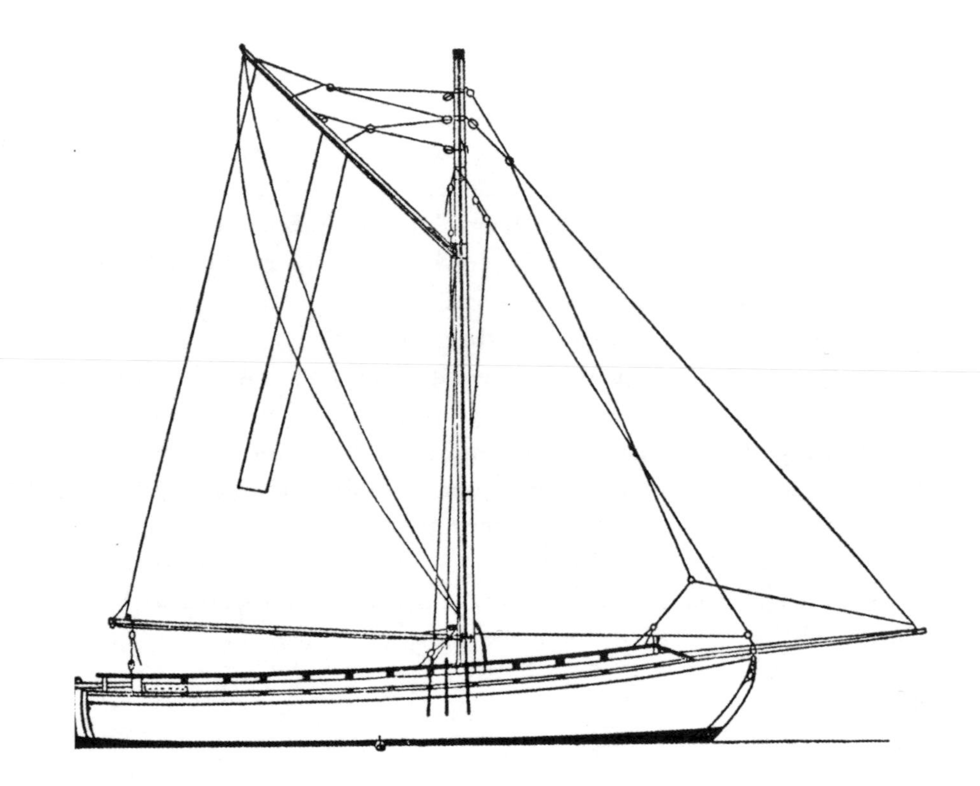

Teddy: Elevation

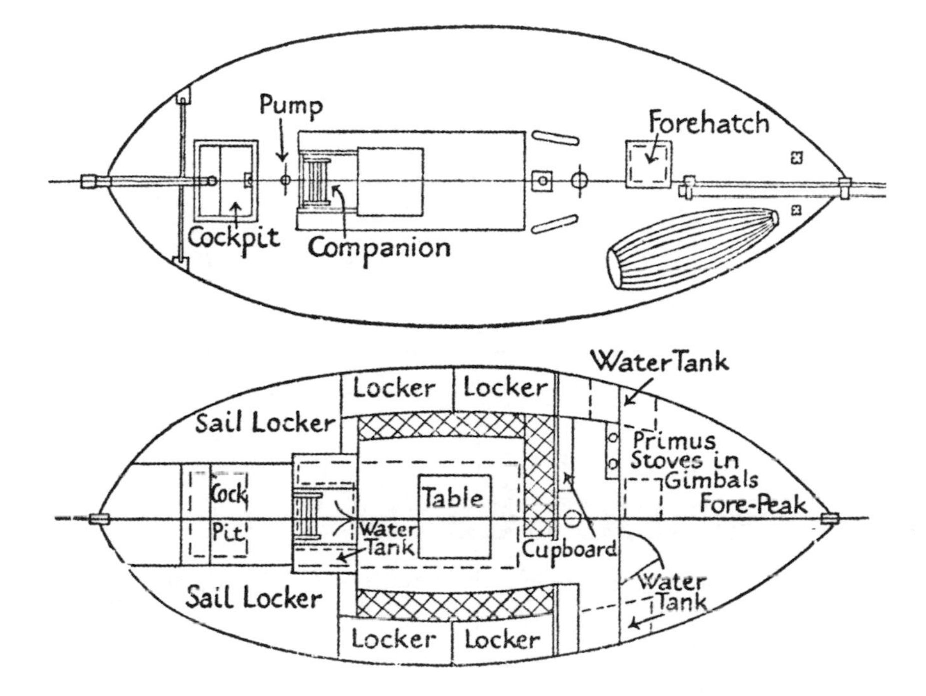

Teddy: Deckplan and accommodation

CLAUDIA MYATT

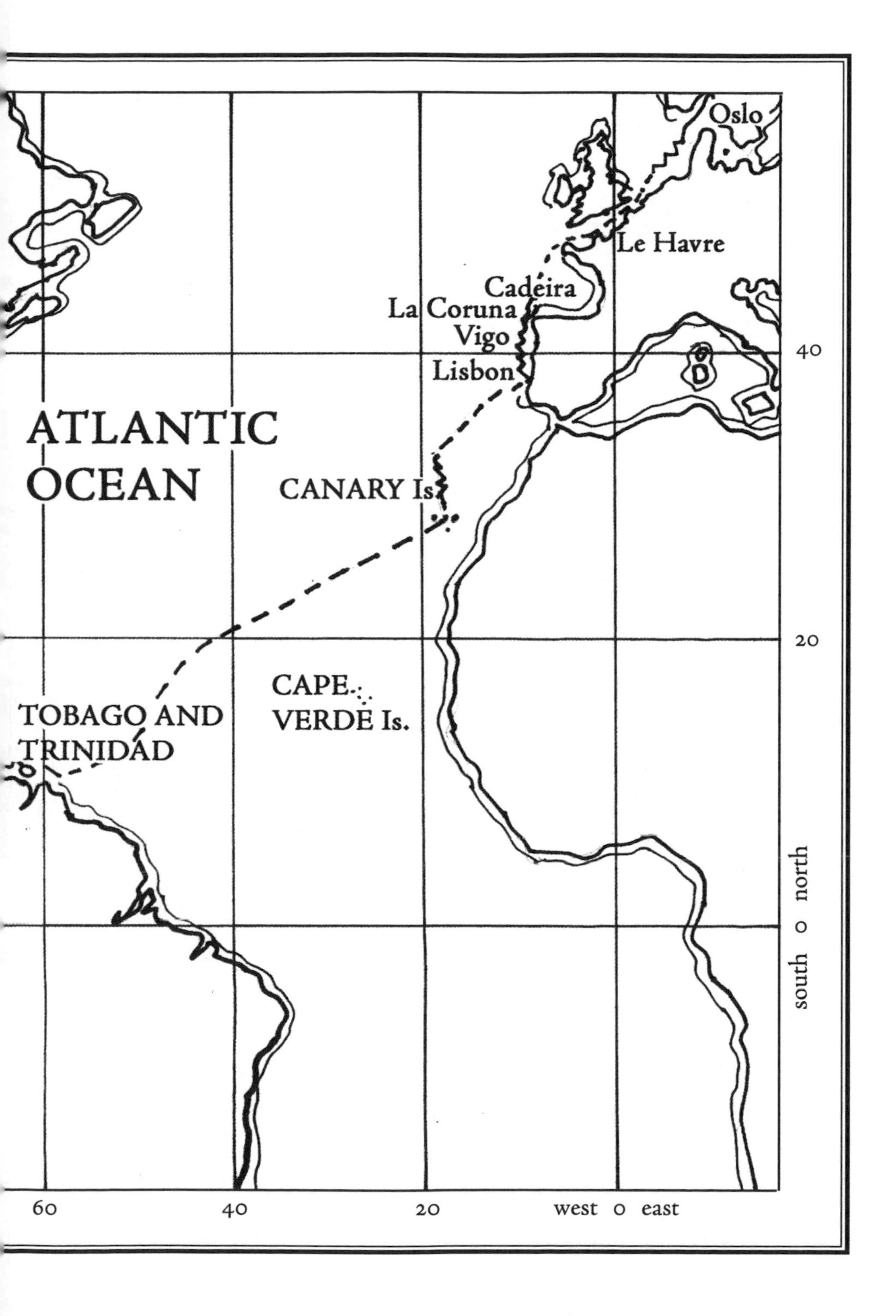
Oslo
Le Havre
Cadeira
La Coruna
Vigo
Lisbon
ATLANTIC
OCEAN
CANARY Is.
CAPE
VERDE Is.
TOBAGO AND
TRINIDAD
40
20
south o north
60
40
20
west o east

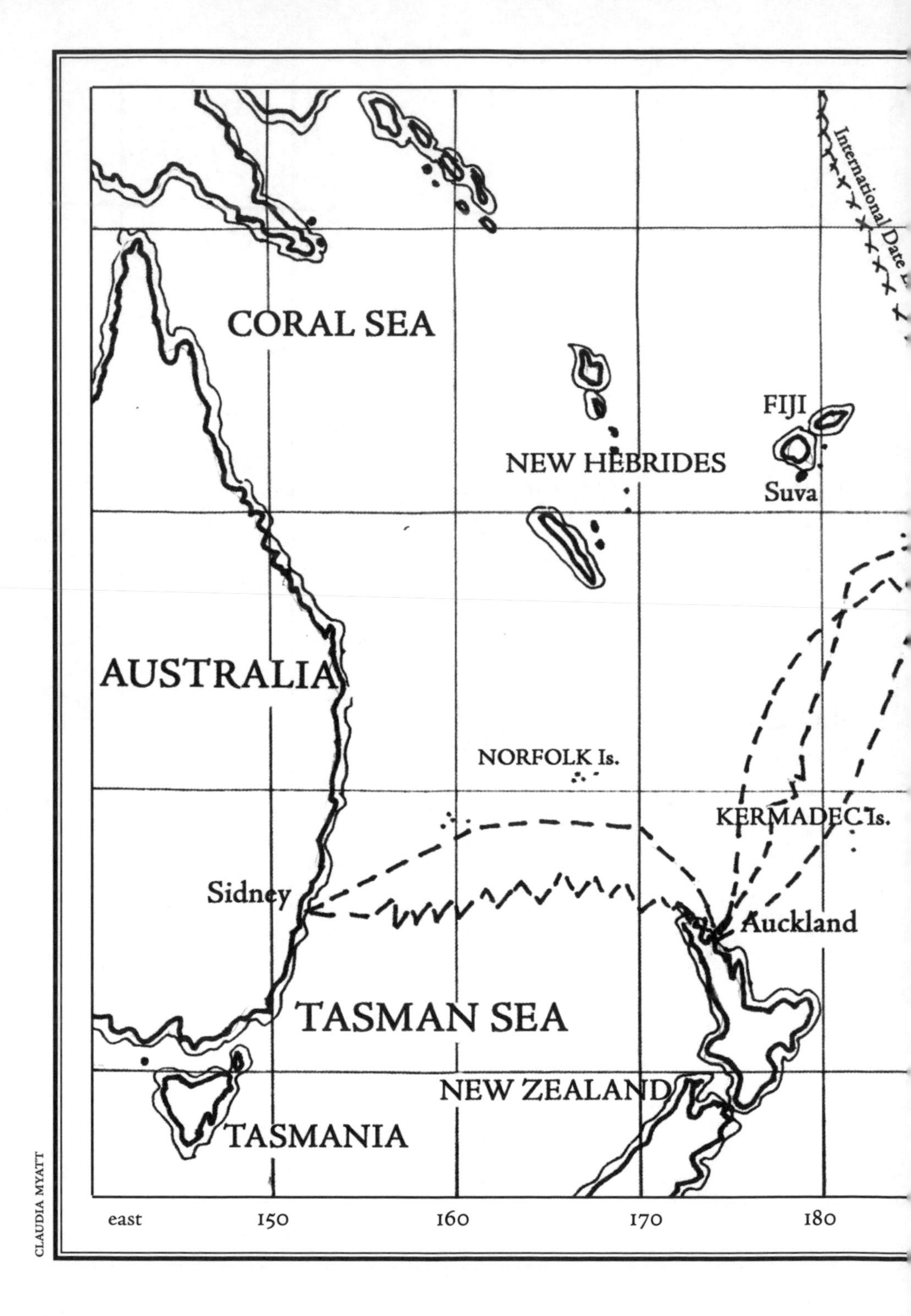
CORAL SEA
International Date L
FIJI
NEW HEBRIDES
Suva
AUSTRALIA
NORFOLK Is.
KERMADEC Is.
Sidney
Auckland
TASMAN SEA
NEW ZEALAND
TASMANIA
east
150
160
170
180
CLAUDIA MYATT

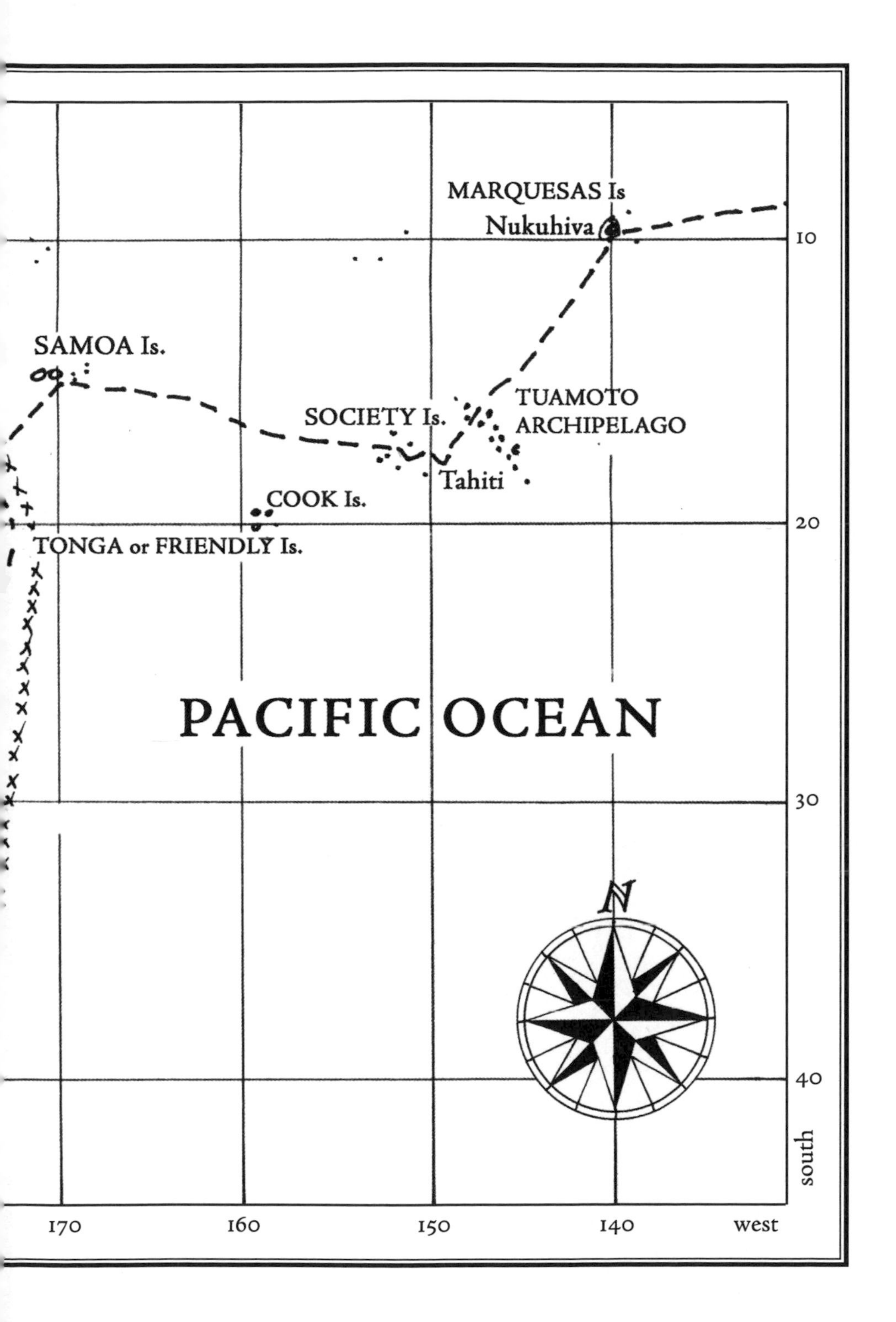
MARQUESAS Is
Nukuhiva
SAMOA Is.
SOCIETY Is.
TUAMOTO
ARCHIPELAGO
Tahiti
COOK Is.
TONGA or FRIENDLY Is.
PACIFIC OCEAN
N
10
20
30
40
south
170
160
150
140
west

How I Acquired Independence

The northern summer was on its decline when we started on our cruise. It was towards the end of August, 1928.

Our home in the years to come was to be a forty-foot cutter built some thirty eight years ago, in my home town, Larvik, by Colin Archer, the famous constructor of Dr. Frithjof Nansen's polar ship *Fram*. The crew consisted of two only, my wife and myself.

Teddy, my ship, had served her day as a pilot boat on the rock-strewn and dangerous Norwegian coast. She had a past—a glorious past, indeed!

I had known her for many years, since the days when I was a small boy and she was the talk of all small boys; when this boat, a wonderful combination of speed and seaworthiness, was the envy of competing pilots unable to boast of a craft so fast and so strong. Many a time did she put to sea through foam and darkness, when winter storms of demonic violence lashed the sea into fury and forced all other boats to seek refuge in port. Surely she had played her part.

Now she had been put out of commission, although she was as hale and sound as ever. But as modern times had made power boats a necessity, *Teddy* had become out of date. This was my chance. I loved the boat and I scorned engines. With the proceeds of my latest novel I bought her. To refit her and purchase the necessary stores and equipment emptied my purse and exhausted my credit; but what did it matter? I had a home now, which, like a Magic Carpet or like an Enchanted Trunk, could take me whither I wanted to go.

I managed to make a contract with a leading Norwegian paper,

Tidens Tegn, which agreed to buy a stipulated number of contributions from me as I sailed along, and even advanced some £60 on this contract. The money all disappeared in the boat, as paint, as tools, as ropes, as canvas. Oh, I wanted a great many things!

Although I supplied my own labour, filling the various roles of carpenter, sail-mender, painter, decorator, expenses of purchase and preparation mounted to nearly £500. By then my resources had run unmistakeably dry, but the old boat was transformed into a trim yacht, and no writ was nailed to her mast.

The inside equipment of *Teddy* was simple; indeed it was very much as it had been when she was a pilot boat. Save for the fresh water tanks, a partition, a few lockers and shelves, and a door leading in to the forepeak, I had not altered her much. The fittings of our *Teddy* were meant for service, not for appearance.

And yet, what with warm colours, cheerful decorations, gay cushions, velvet curtains, and bright metal, the cabin of the old boat was a very cosy place. It was somewhere about nine feet by nine, with plenty of headroom and two built-in bunks on each side. The narrow openings to the bunks were covered with curtains sliding on shining brass rails, as was also the door leading to the galley.

In the galley two single Primus cookers, swinging in gimbals, formed our unpretentious kitchen facilities, which proved nevertheless entirely satisfactory. On each side of the galley was a 110-gallon water tank with taps within convenient reach of the cook. A third tank of the same capacity was placed under the companion floor aft.

Later experience proved that this 330-gallon water supply could, in emergency, be made to last us as long as five months.

The possibility of occasionally replenishing my water supply with rain has often been suggested to me. That idea had indeed prompted me to provide three separate tanks instead of one. So far, however, this scheme has not proved practicable. At times the boat is rolling too much; at others the wind is too high or the sails are a hindrance. Fortunately the problem has never become urgent.

To proceed with our narrative.

Leaving Oslo we followed the south-east coast, calling at our home town, Larvik, to make some final preparations, and later at Arendal, where our dinghy, a 'pram' specially built for me, was awaiting delivery.

When at anchor in this port, and while I was away on some business ashore, the Chief of Police came on board, apparently with instructions from the Marine Department, to prevent us from setting out on what they considered a foolhardy venture. My reception of these disquieting tidings can surely be imagined. My whole heart and my last penny had been devoted to the realization of my lifelong dream: was all doomed to frustration? For one brief moment I even entertained the desperate project of quietly slipping my cable and escaping under cover of night. Recollection of my wife's pledged word that I would not leave without the consent of the authorities rendered this solution impossible. I had to stay.

On the following day, the Chief paid a second visit, accompanied on this day by an expert in the person of the Arendal harbourmaster. After a few pertinent questions, they announced their decision that they considered the *Teddy* unfit for a long voyage. Ah, didn't they know all about that? Hadn't they read the papers? And how could I explain my big fresh water tanks and the various other fittings, indicative surely of my ambitious intentions. Ambitious but foolhardy, they declared, including with a sweeping gesture the lack of spare sails, the absence of nautical instruments, books, tables, charts, etc., etc. In short, owing to the insufficiency of equipment and the shortcomings of the crew, they must condemn my *Teddy* as utterly unfit to cope with the difficulties of an ocean voyage, and they—The Authorities—were there to see that we did not enter upon an adventure that could not but end in disaster.

What could I say, or do? I was no lawyer, and even if I had been, what chance would I have had in a dispute about enactments and amendments with a Government Department and its officials? I explained that I had been trained in square-rigged ships, that I had had experience in sailing small craft single-handed, and that I thought that my seamanship would surely counterbalance my shortcomings as a navigator. Oh, fiddlesticks! Was I conceited enough to think that unaided I could successfully handle a heavy boat like the *Teddy*?

Say what I would, my visitors remained adamant. But suddenly I pricked up my ears. An unwary remark on the part of my main adversary gave me my cue. Though only a vague hint, it nevertheless gave me the impression that the Marine Department, in their instructions to the Police, had admitted the absence of a law which applied to my special case. I decided to make a bold move. I rose. 'Pardon me, gentlemen,' I said, 'but I must get busy.'

'Busy? What with?'

'Weighing my anchor,' I replied casually, 'the wind is fair.'

The two officials looked at each other, then smiled. 'I suppose the game is up,' said the guardian of the law after a while, 'Unfortunately you are right in assuming that we have no legal grounds for stopping you. Although'—he continued musingly—'I dare say that if you would give me some little time to study the issue, I should eventually find a paragraph that would be applicable to your... hm... case.'

'I am afraid that I shall not have time to wait for the outcome of your researches, sir,' I retorted, 'and it would be lamentable if their thoroughness should suffer from the disadvantage of undue haste. You may let me know the result of your legal studies when I come back.'

The old gentleman shook his finger at me. Then he offered his hand and said earnestly: 'Well, I hope that you will come back, and—may neither of you ever repent of this foolishness. Meantime, good luck and God speed you.'

'We have done our best to prevent you from going,' said the harbourmaster, when he shook hands with me, and then, lowering his voice, he startled me by adding rather wistfully, 'but if I were younger, I should have loved to go with you.'

The Authorities departed. The old chief kept waving us farewell, as the launch tore her way towards the town. I felt rather sorry for him. He had not achieved his aim.

We sailed along the coast as far as Ulvoysund, a pretty little cluster of fishermens' white dwellings among the skerries outside Kristiansand.

In the snug shelter of this little cove we passed our last night on Norwegian territory, and thence we set out for the freedom and romance of the deep blue sea.

When the rocky coastline of our Norwegian homeland disappeared behind haze and horizon, my wife in truth shed a tear or two, but for me happiness was unalloyed. It was a glorious morning and the world lay open before us. I felt the cheerful jingle of about one shilling and sevenpence in my pocket—not a large purse, to be sure, to start with on a cruise around the world, but what did that matter? Our sturdy craft was heeling to the wind, responsive and willing, fleeing from care and winter, heading for the sunny land of my dreams.

What need for money? I was a freelancer, dependant for sustenance on whatever loot I could make with my pen. No misgivings felt I about the future; I would assuredly make plenty.

As a side issue, I would call at some of the treasure islands on the way and carry away the riches that my predecessors had left there and that could hardly be of use to them now. Why, money was entirely superfluous at the start!

Would I have changed places with a king? Not I; for I was a king myself. Mine was the staunchest craft, mine was the pluckiest girl, and mine was the utmost degree of independence that mortal man may attain.

I was responsible to no-one. My boat was safe enough for any waters; she would transport me whither I willed. Time and space were at my disposal, seemingly illimitable. I had that feeling of sovereign freedom which fairytales attribute to kings, as the exclusive privilege of princes. Surely there was excellent reason for my high spirits.

True enough, there were a few flaws in our equipment; we had practically no charts; we had neither instruments, nautical books, nor other navigation facilities—only an old card compass, which ran wild whenever there was a bit of sea going. We had no spare sails, and a few other things were missing which to some people would seem indispensable.

But, among the various possessions at our disposal, there figured a bag of potatoes and a fishing line; thus our fare appeared well secured.

The boat was in good condition; likewise her sails and her rigging. Her white-enamelled sides glistened wet and smooth like polished ivory. The heavy canvas of her sails bulged evenly in the strong and steady breeze. She was a beauty!

2

A Rough Passage across the North Sea

The fine weather experienced at the start was short-lived. On the very first night a hard blow sprang up, and thenceforward we had nothing but gales until, sixteen days later, we made the port of Le Havre, somewhat the worse for wear, but happy in the conviction that our confidence in the boat was justified.

To be sure, it was too heavy a craft for one man to handle, and more than once the tasks I had to tackle almost broke me. A strenuous job reefing those heavy sails in the midst of a frenzied gale, while the boat is being tossed about by a raging sea! Very often was this reefing necessary at night, when nothing was discernible through the pitchy darkness but flying spray in the dim red or green glow of the side lights and the ghostly phosphorescence of the tremendous combers; when nothing could be heard but the roar of the tempest, the rattle of the gear, and the violent flapping of the furious canvas.

I had to shout my directions into the ear of my wife, who stood at the tiller drenched and frozen, faithfully sticking to her post. Unable to see me through the darkness, she would call my name from time to time; but her voice was lost in the storm and she received no answer but the shrieking of the gale. And then she would not know if I were still aboard. Assuredly she suffered some hardships!

I shall relate one outstanding experience of this kind:

Crossing the North Sea it had been my intention to pick up Terschelling lightship off the Dutch coast. However, owing to continuous beating against south-west gales, we made a landfall some thirty miles east. It was on the night of the ninth day out, when we came within range of the first light, which, according to my dead reckoning, should be Amelander Gat.

I had no list of the lights along the Dutch coast or, for that matter, of any other coast in the world, and the ancient North Sea chart on which I had been zig-zagging our changing course did not give the correct characteristics of the lights. Presumably they had been altered many times since those remote days when my chart was compiled. This I knew, and when, therefore, the flashes which I took to be from Amelander Gat did not correspond with the characteristics given in the chart, I was not greatly disturbed.

During the evening the wind had abated to a stiff breeze and, having at the same time pulled round to south-east, it was fair for the first time in nine days. Moreover, as the wind blew offshore, the sea soon became comparatively smooth. Therefore I shook out a reef and shaped a course parallel with the coast. With the breeze on the beam we were soon travelling at some nine knots through the night.

It had been my hope that one or the other coastal lights might still flash as stated in my chart, but although during the first half of the night we observed three or four other lights, not one of them would fit in.

Meantime the wind had again veered round to south-west, at the same time increasing steadily. The night was pitchy dark. It looked as though we were in for another gale. At midnight I had to stow the jib. Two hours later I took two reefs in the mainsail and one in the staysail, but while I was at this work, the force of the gale rose so rapidly, that even before I tied the last reef point it was quite clear to me that we had far too much canvas spread. Nevertheless, I decided to wait till daybreak before further reefing down. I was exhausted with battling against that heavy canvas, while the little ship was heaving and plunging, and the idea of tackling another reefing job in the blackness of that night seemed like a nightmare to me. Yet those endless hours of tossing about in a full storm under that terrible press of canvas were enough of a nightmare in themselves.

At last dawn came, a wild dawn to be sure, but still a dawn and after such a night even the most cheerless daybreak is a relief.

Then out of the flying mist appeared the stately shape of a 10,000-ton steamship. She was in ballast, heading almost straight for us and steering a north-north-east course. Here was an opportunity for ascertaining our position; wet, frozen, and deadly tired, we both longed for a rest. If I could only make sure where we were, we might perhaps make port somewhere on the coast.

So far we had known nothing but continuous interruptions even when, off watch, we had been trying to snatch a weary wink in our oilskins on the hard and wet bottom of the cockpit. We had not known dry clothing for days. All our working clothes lay in huge soaking heaps on the cabin floor.

Naturally, therefore, the thought of a night's undisturbed rest, the thought of dry clothing and of a good meal, spurred me to renewed efforts. Our flag halyards had carried away with the chafing of those persistent gales, but in spite of the terrific heaving of the boat, I managed to climb the rigging and display the flag signals: 'Please give us the distance and bearing to Terschelling lightship.'

The steamer passed us within half a cable and must have seen our signals, and furthermore recognized that we were in trouble, but callously pretended not to notice and our appeal remained unanswered. To her this short choppy offshore sea, worse than most big seas to a boat like ours, was a mere ripple. Stately and on an even keel she pursued her course unperturbed, and wearily I clambered down.

Whilst the steamer was approaching us, my wife was full of gay expectancy. The dejected expression her face now bore made me feel truly sorry for her. But at the same time I felt that the cavalier attitude of that master or his officers was not only a slight upon a fellow seaman in a small ship, but also an insult upon the best and finest traditions of the sea.

'Never mind!' I consoled my wife, 'We shall find the way for ourselves.'

By approaching the coast I trusted to come across a lightship—the name of such ships is always painted on the sides in huge letters—or we might meet some other ship, which might give me our position. I therefore put *Teddy* about, but I did not reef her down. The weather seemed to be clearing up. Sometimes even the sun showed his bleak, half veiled face between the hurrying clouds, and with the wind south-west—off-shore—we should soon be in smoother waters. Moreover, in view of the probability that we should have to combat the sweeping tidal currents inshore, and perhaps in some entrance, I thought it advisable to keep as much canvas on her as she could safely carry. If I had possessed a barometer I might have acted otherwise. I still think that if I were again to cross the North Sea under similar conditions, and were given the choice between these two instruments only, I would rather have a barometer than a compass.

An hour's hard sailing through tumbling seas brought us within sight of the low land. Water towers and larger buildings stood clearly out against the grey of the southern sky. No lightships, no steamers were in sight. Yet I went on, hoping that some large signboard, a factory ashore, or something similar, might give me a fair idea of our whereabouts. Thus we continued in ever smoother water until we were within perhaps four miles of the land, when all of a sudden, in a squall, the wind veered round to west-north-west and, gaining velocity from minute to minute until it piped through the rigging with hurricane force, sent us flying along through sheets of foam and spray. In less than a quarter of an hour the storm, now blowing in from the sea, tore up rough waves, and it was all I could do to manage the tiller. The waters off the Dutch coast are shallow a long way out, and the waves, rapidly growing with the tremendous force of the gale, soon turned into hollow breakers, yellow with sand from the bottom. I was compelled to put about. There would not be time to reef her down on this tack. It would be a trying ordeal for the sails and rigging, but it had to be faced. I shoved the helm hard down. She responded readily, as always. The canvas flapped furiously, shaking the whole boat as if she were in convulsions. The rattle of the gear was terrible, it was indeed a marvel that the rigging stood up to the strain. But then, when she was nearly head to wind, a comber caught her under the nose and beat her off, setting her racing at a frightful speed towards the shore, until I could bring her to the wind again. I could not wear her without lowering the mainsail first; a jibe all standing would decidedly have pulled the rigging out of her, and it would have taken a long time to clear the tangled mess into which the breakers had converted our running gear. Moreover, I could not leave the tiller. And yet I must bring the boat about.

I tried time after time; always she was headed off by those ugly seas and then she would resume her mad race towards disaster. We might expect to strike at any minute, and then it would be the end.

My wife stood by my side in the cockpit, helping me to handle the sheets, her nails all torn and bleeding. She saw the danger we were in; she could not help seeing it, but uttered no word of complaint.

It seemed all up. The boat was again falling off after one more futile attempt to put her about, and the roar of the breakers on the nearby shore was distinctly audible as an undertone to the shrieking tempest, when my wife suddenly put her arms around my neck and, kissing me through spray and tears, shouted into my ear: 'It does not matter, Erling, as long as we are together.'

Believe me, it was good to hear that.

And then the wonder happened. I had again put down the helm mechanically, and the boat plunged up and down, shipping heavy seas over her bows, while the terrific rattle of the heavy mainsheet blocks and the gunfire of the shaking canvas seemed to drown even the infernal concert of the gale and surf, when suddenly the rattle ceased; the mainsail had filled and the boat heeled over on her port tack, heading out to sea.

It was a narrow escape, but when, two hours later, I had taken in the last reefs and hove to, we were out of danger. We were well away from the shore and the boat rode over the huge combers like a duck, hardly ever wetting her deck save for spray and an occasional squall of rain. Still the storm raged with undiminished force. I watched her, until I was satisfied that I was quite superfluous at the tiller, which I had pegged down and lashed. I was just about to go below, when, out from the north-north-east, came, heaving and staggering like a drunken man, the huge black hull of a steamer in ballast.

It was our friend of the morning, who was running back to port and safety. Again she passed us at half a cable's distance, but now it was her turn to be in trouble. She rolled so frightfully that I could have counted the planks in her bridge deck whenever she tumbled over towards us. *Teddy* lay snug and safe, mounting with ease the seas that seemed to worry that steamer so much.

Although it was still early in the afternoon, I presently lit the side lights and went to sleep, and for fourteen hours neither of us stirred. We slept dreamless and fast until the next morning, while the worst storm that I had experienced for many years was raging. We did not keep a look out, though we were in waters where usually the traffic was so dense, but at that particular time we seemed to have the sea all to ourselves.

I had no fears of being run down. The chances of meeting a sailing vessel hove to on the other tack were negligible, and according to the rules of navigation all other ships had to steer clear of us.

A few days later, when sailing in the English Channel, I found, though, that this rule of courtesy to sailing ships without auxiliary power is often disregarded, at least in the daytime. The reason is, I presume, that steamers, even when meeting a small vessel under canvas, expect that vessel to be equipped with an auxiliary engine, in which case she would have to observe the same rules as a steamer. Wind-jammers and other honest unadulterated sailing craft are steadily becoming scarcer everywhere and are rarely seen on the high seas. At night, of course, the absence of a masthead light on a sailing vessel shows her nature to meeting ships, but even then I have occasionally found that in the rush of our times, some people are liable to forget the rules of the road as far as sailing craft are concerned. They seem to be all too cocksure that a sailing vessel cannot travel as fast as they are travelling, no matter what kind of a smoking coal-barge they have under their feet.

We were, as I have said, in the Channel. Double-reefed in a moderate north-westerly gale, we were flying along in great style, when my wife, who was at the tiller, observed a tramp steamer on our port bow. I was just about to prepare our Sunday breakfast, which I intended to turn into a proper grand affair, including mackerel caught by an energetic wife at daybreak, potatoes, bacon, eggs and good strong coffee.

Before properly settling down to my job, I went and had a look at that steamer. She appeared to be steering a course at right angles to ours but, as we were some four miles away and the visibility was good, I decided that it was alright and went about my business in the galley. That collier seemed to worry my wife; apparently I had not quite succeeded in making her believe in that all too idealistic rule of the chivalrous social equality of ships at sea, and she called out several times, but I only shouted in return: 'Keep your course!'

I had removed one of the Primus cookers from its gimbal and placed it on the floor, so as to furnish the frying-pan with its full measure of flame. Steadying the pan with one hand, while keeping a vigilant eye on a range of pots laden with agreeable odoriferous culinary achievements, I was kept very busy.

But then my wife called out again, and her voice had an insistent note of anxiety.

I rushed on deck and grabbed the tiller.

That skipper, instead of passing astern of us as he should have done, and surely could have done without going ten yards out of his way, had altered his course three or four points in an effort to pinch across our bows. Now we were wedging together alarmingly near, and although *Teddy* made better headway than that tramp, the odds were that our stern would be caught by her bows if we kept our course. Even then it was the skipper's duty to back his engines and let us pass, but, seeing the utterly helpless expression on his face as he stood there on the bridge, I let our boat spin into the wind and went clear. Then I kept her a-shiver until the steamer's ugly stern had passed us.

Hauling the windward staysail sheet to come away again, I saw her log line. I would have carried it away if I had proceeded on our course then, so I put her into the wind a second time to spare that log. This considerate action I immediately regretted, since at this stage we were given the benefit of her huge stern waves—that wretched tramp dragged half the ocean behind her—and *Teddy* with her sails all a-shiver began to roll in frightful fashion. An ominous crash and rumble from below told its own tale—the Primus with the pan, hurriedly propped up against a table leg, had capsized.

Incensed, I shouted well-merited abuse, shaking my fist at the cause of it all, but that skipper, obviously misinterpreting my action as a gesture of friendly farewell, waved back most graciously. I turned away in disgust and we proceeded on our course.

But alas for my odoriferous culinary triumphs! And alas for anticipated pleasures of the palate! There, hopelessly ruined, sliding in one greasy mess from side to side of the galley floor, lay our great Sunday breakfast, sublime product of inborn culinary talents and acquired skill.

We had had another close shave on that passage a day or two before.

It was about midnight. A stiff north-westerly was blowing. We were sailing hard and were approaching Haak lightship at great speed. Anxious to get the best possible check-up on my dead reckoning and also desirous to show my wife what a lightship looked like, I kept very close.

The flashes at that range were, naturally, very blinding, so that I could not see the compass when I looked at it. Therefore, I kept my eyes upon the black space ahead to re-accustom them to the darkness.

When the next flash came, it revealed, not ten yards ahead, the huge black mass of one of the vessel's anchor buoys. It shone white in the flooding light. We just managed to swerve clear. But what a marvellous stroke of good luck!

3

We Make the Acquaintance of Rich Men ~ A Friendly Spirit
A Treacherous Port and Strange Marine Edibles

We stayed at Le Havre for three weeks, and took full advantage of the rest we so much needed after the strenuous time we had had at the outset.

A leading paper printed an article about the *Teddy* under the title 'La Belle Aventure.' It was as though written out of my own heart; my own visions, my own dreams put into simple and beautiful words. That writer made a friend of me. There are people in every country who feel as I do and who might want to do the very same thing which I was doing if they could only break away from their surroundings. But that is not so easy.

We were moored in the Bassin de Commerce, in the very heart of the city, amongst a number of the finest yachts in the world. A few yards away lay—like a palace of gold and ivory—Baron Rothschild's exquisite steam yacht. When we went ashore, we would leave our dinghy at the steps right under her stern.

One day, on our returning from a stroll in town, a fine old gentleman, whom I suspected to be the great financier himself, spoke to me from the yacht. He had seen some small boys playing around my dinghy, he said, and had had them chased away.

Subsequently I used to amuse myself, when disembarking, by telling my wife: 'Baron Rothschild will look after the dinghy.'

Everybody was friendly to us and interested in our venture. Those three weeks in Le Havre have become for both of us a series of happy reflections. Among the numerous visitors who came to see the boat was a British colonel. He wanted to buy *Teddy*.

'I love your boat,' he said.

'So do I,' I responded.

Then followed an offer amounting to three times the price I had paid, including repairs and outfitting.

No response.

'I am a very rich man,' observed the colonel with considerable dignity.

'Not rich enough to buy this boat!'

That concluded the interview.

But day after day the colonel returned to the quay. He, too, had fallen in love with our boat and he coveted her.

But how could he ever dream that I would part with my kingdom for mere golden guineas?

His longing went ungratified.

And yet I could hardly say that I was rolling in money at that period.

From Le Havre our path often lay along well-beaten steamship tracks, the wind being mostly fair, and thus allowing us to pursue such a course. Of the steamers we passed day and night, some blow their whistles, some dip their flags to greet the little boat sailing so bravely in spite of the rough weather and high seas of the Bay of Biscay.

Though oftentimes I longed to heave to, we kept her going, for the autumn was well advanced and it was advisable to go south. On one occasion only was I forced to lay her to for twenty-four hours. A heavy gale was blowing and the fore staysail had carried away.

It was the night following this storm. The wind had abated somewhat, but I still kept her under close-reefed canvas, waiting for the sea to moderate. Moreover, I had to mend the fore staysail before we could carry on, and for this work daylight and dry weather were necessary. Meanwhile I had slackened the sheets an inch or two, so that the boat made a couple of miles headway while steering herself on a southerly course.

I took the watch from two o'clock and was sitting in the cabin smoking my pipe. My wife slept in her bunk. From time to time I would put up my head from the companionway and take a look out. Nothing to see but scudding clouds, breaking seas, and now and then a glimpse of waning moon.

In a sailing ship, in wind and sea, a variety of noises form a continuous, monotonous concert to which one gradually becomes accustomed. There is squeaking and creaking, knocking and rumbling, groaning and shrieking. Such sounds contain no disturbing elements. One accepts them as natural—learns in time to distinguish and localise each noise. Eventually the melody of sound grows soporific.

I must have dropped asleep on the bench when suddenly I awoke. I had heard a foreign noise, a knocking I could not locate. However, being very tired I must have fallen asleep again, when with a start I was jerked into complete wakefulness. I heard again the unfamiliar sound, a distant knocking, three times repeated. I listened, sitting bolt upright. For the third time came the knocking—insistent, warning, as if produced with a hard knuckle—and the sound seemed to issue from the companion top.

Seized by an instinctive uneasiness, I rushed on deck. Ho! What was that? A red light close ahead on the lee bow, and under it the huge black hull of a wind-jammer on her starboard tack.

Hard down the helm! Will *Teddy* come over? A tense moment! The huge bulk of the barque comes nearer. It rises and sinks in the sea while the water shoots out through scuppers and weather ports. The fore staysail comes aback. *Teddy* falls away. We go clear.

The barque sails by. A man is standing by the poop-rail. A quiet voice says in French, 'You have good luck, Monsieur!'

This episode I did not relate to my wife until a year later.

Of course there is a natural explanation; there always is; even though I cannot in this instance find one in which I have any confidence. Yet it may have been the fore staysail flapping, in which case the sheet rope, which runs along the cabin top, might produce a rapping sound.

Why should this happen just as that critical moment, and neither before nor after?

As a partly serious explanation, I assigned the credit of the warning to the benevolent activity of the ship's guiding spirit, whom I named after the old pilot for whom the boat was built and who subsequently met his death on board. It is to his tutelary agency that I have ever since attributed the many strange happenings that have occurred on the *Teddy*.

After leaving Le Havre we did not sight any land until, at the end of a nine days' voyage, we made out the huge black masses of Cape Ortegal, on the north-west coast of Spain. The following day we cast anchor in pretty Cedeira. It had been our intention to make for La Corunna, the principle town on this part of the Spanish coast, but a howling gale from the south-west made beating to windward against the current both wet and miserable, and prompted us to seek shelter where we could make it.

The entrance to Cedeira is narrow and rock bordered, and fairly dangerous to attempt without a chart and local knowledge. We managed, however, to pick our way into the bay, beating against strong gusts of wet wind which blew out through the gap. Turning to port around a point on which stood the ruins of an old fortress, once the stronghold of pirates and smugglers, we came to anchor outside a fleet of fishing boats in four fathoms of water. The sea bottom consisted of fine hard sand.

It being early in the day, we decided to go ashore. Some fishermen showed us the way to the village. A series of rough, well-worn steps, hewn out of rock, led to a narrow footpath winding along the steep edge of the cliff and sometimes overhanging it, where the rock beneath had tumbled down. Gradually the path widened; an ox cart with big rough wooden wheels rumbled towards us, a discontented donkey passed us, carrying heavy burdens on his sides, an old woman went by with a huge load of firewood on her head, bigger and probably heavier than herself. 'Buenos días,' she said, in passing.

As everyone we met repeated this greeting, we promptly adopted the custom. Then all at once, as we reached the slope of the hill Cedeira extended before us, a most picturesque cluster of buildings that looked and probably was hundreds of years old. At least the sanitary arrangements of the village had certainly not improved much since the middle ages.

Wading through mud we approached the town only to find the access barred by a huge war-like sow with some ten young suckers. My wife flatly refused to brave that formidable foe until a small boy, kicking the sow with his grimy bare feet and pulling her tail, marched her and her numerous progeny off into his own family's living room. Domestic animals seemed to enjoy distinct privileges in Cedeira: pigs, dogs, fowls and monkeys were walking in and out of the houses as if that were the most natural thing to do.

We passed through narrow cobbled lanes, winding between white-washed walls of ancient and uneven masonry. Most of the houses were two-storeyed, the ground floors windowless and the top storey supporting rickety wooden balconies almost touching one another across the street. A flight of deeply worn stone steps led down to the bank of the little river, beyond which lay a wide clean sunny beach.

We did not linger long. There was not much more to see. Besides, the wind was chilly. The sun had disappeared and with it much of the charm of that queer little place. It was good to come back to *Teddy* and our own modest conveniences.

The next morning I noticed that the gale had shifted to north-west and, blowing right in through the gap, had brought up some sea in the bay. The fishermen, anchored inside of us, were all getting up steam. At the time I did not realise the full significance of this measure, although even then it was obvious that they could surely not be going to sea in that weather.

But soon the wind began to blow in fierce squalls which rapidly converted the entrance to the bay into one mass of tumbling roaring foam, and sent clouds of spray and lather flying a mile or more across the bay. There was no hope of getting out of the harbour; we were all caught as in a trap.

Since the hard sandy bottom offered but poor holding ground for our anchors, I took the precaution of making the sails ready, reefing them down to the last point. In addition I put a good solid buoy on our anchor chains in case I should have to slip them.

It was none too soon. I had barely finished my preparations, when the anchors broke loose and we went tearing down towards the rocks as though those hooks of ours offered no resistance whatever. Meantime, everything being ready, the sails went up in a hurry and my wife took the tiller. Then, tacking to windward, I managed to pick up the hooks and, assisted by some fishermen, who brought me a man-sized grapnel, we sailed the boat back to her anchorage. There, dropping each of the three hooks in a carefully selected position, we considered ourselves very clever and the boat safe. Which proved a sad mistake.

A dozen times or more we dragged our anchors and had to slip them and go for enforced sails.

Of course this always happened when the tempest had gathered up its forces to their topmost pitch, when the gale was piping in its highest key and when visibility was at its minimum.

Night and day I was on the lookout with every nerve strung to catch the first sign that we were again dragging; and when those squalls came on at night with impenetrable darkness, or with driving rain that blotted out every contour visible by day, how often did I wish that we had been well out at sea.

Yes, indeed, that bay provided us with many an exciting sail, while tense we listened for the breakers and strained our eyes to capture the dim outlines of a near shore.

However, after a few days of this unenviable experience, the skipper of a schooner, taking pity on us, sent me his warp anchor, a 500-pound affair, and thereafter we lay safe.

Those weather-bound fishermen in Cedeira taught us to eat barnacles and octopus, which, after overcoming our first foolish aversion, we found delicious fare. The barnacles grew in big clusters along the rocky shores and were gathered at low tide. They were easy to clean, and when boiled in salt water their long necks tasted very much like shrimps.

Maybe these barnacles were of a different kind from those which had the annoying habit of settling on the copper-sheathed bottom of our boat at sea. At any rate, I have never had the courage to eat of our own crop, and I have certainly never felt inclined to cultivate them.

So far, Spain is the only country I have visited where these shellfish are eaten. The octopus, however, seems to furnish an essential ingredient in a variety of delicacies used in many of the tropical or semi-tropical places which we have since visited.

4

We come to La Corunna, Meet Jesus and have Experience with Burglars, Authorities, and Popularity

The storm continued for nine days and when at last it abated we sailed to La Corunna.

La Corunna is an interesting town. Possibly founded by the Phoenicians some three thousand years ago, it has played a part in history on many occasions. It boasts of buildings representing various historical epochs, some of these buildings being of great age. For instance, the Hercules Tower, more than three hundred feet high, dates back to the Roman period, perhaps even before that, and has since been in the possession of many nations, at the same time a stronghold and a beacon, until today it serves the sole, peaceful purpose of a lighthouse.

We found the Spanish a charming people, kind-hearted and congenial. Wheresoever on this cruise I encountered Spaniards I found willing hands and a ready smile.

Only one unpleasant incident befell us in La Corunna, but even that now appears to me to have been a blessing in disguise, since it brought us a wholly undeserved popularity and the sympathy of, I might say, the entire population.

Setting out on a voyage which would in all probability have a duration of several years, we had taken with us from Norway our entire wardrobe, much to my subsequent annoyance. It occupied too much room, it demanded a great deal of care, and yet, whatever attention was bestowed on it, its ultimate ruin was a certainty. To air and brush mildew off evening clothes on the high seas was a hopeless business, which nevertheless seemed to thrust itself upon me most unpleasantly often.

When, therefore, an enterprising thief in La Corunna broke into the cabin during our absence and carried away practically all our clothing, I soon found consolation in the thought of the work and trouble I should be spared; for, in the end, clothes that cannot be of immediate use are only an encumbrance.

The burglar had not taken away our oilskins and sea boots; neither had he taken much of our underwear; and thus if he had only left us some of the old overcoats which we used at sea to keep us warm, I should have felt nothing but relief at our loss.

Among the people who befriended us in La Corunna was an old boatman, called Jesus, a name which we discovered to be common amongst the Spanish. I was greatly taken with this appellation, for I can henceforth speak most truthfully and proudly of my friend Jesus. He was a very good fellow and an excellent sailorman, always ready to give me a helping hand when I wanted one, whether for odd jobs about the boat or for shifting our moorings, which very often became necessary on account of the heavy weather which prevailed during our three weeks' stay in La Corunna.

The credit is also due mostly to him that people showered so much kindness and sympathy upon us.

My knowledge of Spanish was confined to two or three score words; his English had similar limitations; and yet by gesticulating, drawing pictures and kindred means, we carried on long conversations, sometimes to the undistinguished mirth of occasional onlookers. At times I felt sure that Jesus did not understand me, but Jesus always understood, or if he did not he would put his own interpretations on my remarks, and I have reason to believe that his interpretation would generally be undeservedly flattering to me.

It was he who set things humming when I told him about the burglary.

When we returned that night to the boat and discovered the robbery, my wife—grieved at the loss of her entire wardrobe—insisted on my interviewing the police immediately and bringing an officer to the scene. He came to the pier but would not go aboard; the surging waters did not appeal to him.

Accordingly I tried to explain the position to him then and there, using the English language as the medium most likely to be understood. When I had exhausted my eloquence, he said: 'No comprendo Noruego. Parlez-vous Francais?'

Whereupon I told the story all over again in my best French. He listened with polite attention till I finished, then quietly remarked: 'No comprendo Noruego. Parlez-vous Francais?'

I repeated the tale a third time in French, speaking very slowly and inserting a Spanish phrase or word whenever possible. He listened very carefully and I even thought that I saw him nod once or twice, which encouraged me to go on, to repeat and to explain until I was exhausted. I stopped and looked at him expectantly. He threw out his hands and again began his: 'No comprendo Noruego. Parlez...' But I had turned away in exasperation and wished him, 'Buenas noches!'

Which he understood.

I would have let it rest at that but my friend Jesus would not hear of it. He alarmed not only the Commandanzia, the Capitano del Puerte and the police but also the newspapers, and presently photographers and reporters swarmed aboard.

Two days later, *Teddy*'s picture and ours appeared on the front page of a dozen papers all over Spain under headlines which, I fear, were more flattering than true. What appeared to afford particular amusement was my secret rejoicing at the loss of that evening suit, whilst outwardly I had to wear a mask of sorrow to please my wife.

This I had confided to no one but Jesus, who had, however, passed the story on to the papers with his own embellishments. Unfortunately my wife saw some of those papers and expressed to me freely her opinion.

Many of the resident people deplored the fact that the incident would be certain to lower my opinion of the Spanish character, but to all and sundry who approached me on this subject I said: 'Malos hombres en todas partes,' which pleased me immensely, partly because it expressed my definite conviction, and partly because it afforded a unique opportunity for expressing a conviction in a perfectly good Spanish sentence.

The remainder of our stay in La Corunna brought great fun. Bunches of young and pretty girls brought us big bouquets of flowers with ribbons in Spanish and Norwegian colours. Young boys rejoiced to run errands for us and fiercely declined the money which I offered them in return. A gentleman whom we never met sent us a Norwegian newspaper every day or whenever the mail arrived. Other unknown friends sent us chocolates, fruit and more flowers. The Authorities were friendliness personified, and when at last we left on a bright Sunday morning all La Corunna seemed to be on the quays to wave us farewell.

On that occasion I nearly made a fatal blunder.

Just off the main quay a big British schooner-yacht lay at anchor in the stream. She had come in for shelter and repairs. Attempting to beat around Cape Ushant on her way to Southampton she had lost all her sails and finally had to run before the gale right across the Bay of Biscay to La Corunna, where she had arrived with skylights shifted, bulwarks smashed, and her long overhanging counter looking very much like masonry after an earthquake. From this, even at the risk of digression, a moral may well be extracted—for such as would go ocean cruising, the slogan: 'Short ends for safety; they lift!'

All the camera owners in the country appeared to be assembled on the quays to snapshot *Teddy*. Seeing no reason why I should disappoint them, I took a short sail along the wharf and then, intending to sail away inside the yacht and across her bows, I eased off my sheets and put up the helm. Suddenly I discovered that a very strong current was carrying us rapidly down towards the yacht's long jib boom, where I should certainly have hung up the boat and made a terrible mess of her rigging, if I had not immediately brought her to again.

We managed to scrape past the boom end with half an inch to spare and were greeted with applause by the public who seemed to think that I was doing tricks—a nautical looping the loop—for their amusement, and who must have thought me in possession of a truly wonderful judgement.

They might not have thought quite so much of me if a merciful Providence had not provided that half inch of space between our shrouds and the end of that jib boom. So narrow is often the division between fame and infamy, between fortune and disaster.

Outside the peninsula, which forms the harbour of La Corunna, a mile or so north of the Hercules Tower, appears an ugly rocky patch, on which the sea breaks heavily. To my dismay the breeze died out when we were close to this patch, and I experienced an anxious time when I saw the tide setting us towards it. Luckily, however, a strong southerly wind came up before it was too late and soon carried us clear of danger.

Obviously, fine weather was not for us, though. The same wind soon developed into a gale.

Beating against wind and current it took us fully twenty-four hours to come abreast of Cape Finisterre, and there our progress stopped. Under close-reefed canvas it was impossible to advance noticeably against the northerly set.

I had been hoping to make Corcubion, a harbour just south of the cape, but as I had no chart and further had been told that Corcubion is surrounded by dangerous rocks, it would have been necessary to make the entrance before dark. Seeing that this was out of the question, and rather than heave to and be driven back by the current, we decided to make for Corme, which had been described to us as a good harbour and easy to enter.

We wore ship off Finisterre at 1.20pm. and dropped anchor at four o'clock sharp having travelled a distance of thirty-three miles in two hours and forty minutes, certainly no everyday performance for a small boat.

We passed a trawler on the way in; we had to stow the staysail and lower the main peak in order to get a chance of enquiring about the anchorage at Corme, and these trawlers rarely do less than ten knots.

We let go our anchor in the midst of a squall. It looked as though we had dropped it altogether, so fast did we drag. With fifty fathoms of chain and a 180lb grapnel at the end of it, we travelled quick and lively inshore towards the rocks. It looked critical, but fortunately some fishermen in their dories had also been caught in that squall and were hanging on to the anchor chains of an old steamer, laid up close inshore apparently for ever. I managed to throw them a new 2½-inch line as we tore past. They made fast to one of the anchor chains and I succeeded in checking *Teddy* without breaking the line. It was our last chance. Fifty yards further in were the rocks.

We moored alongside the steamer. It was by no means a comfortable berth, as we were rolling and bumping against her sides continually. However, about six or seven o'clock that same night the harbour pilot came on board in a staunch whale-boat with four oarsmen. With considerable difficulty these contrived to make fast with a 4-inch line onto a heavy mooring buoy in the harbour, and by joint forces, using the staysail to assist us, we finally brought her close up to the buoy and moored her with chains and wire rope.

The next day we had all the local authorities on board: many bringing flowers and souvenirs. They were simple and charming people. As the boat was heaving and rolling all the time, however, the majority of our visitors made only a brief stay.

Induced to go ashore the following morning, we found that these kindly people had prepared for us a real reception. Since the school had been given a holiday of the occasion, we had a numerous escort wherever we went.

Corme is picturesque but poor. Fishing and fish canning seemed to be their chief, if not the only, source of revenue. And yet the generosity of these impecunious people became almost embarrassing.

My wife wanted to buy some fruit and vegetables. A pair of young school mistresses wedged in and paid for them.

We had also run short of coffee on the boat and had it in our minds to buy some, but seeing that no one would accept our money we had to refrain from mentioning the word, and therefore we went to sea without coffee, it is true, but with very happy memories.

5

Sailing on a Friday
Enter 'Spare Provisions' ~ Discoveries with a Sextant
Madeira, the Canary Islands, and the Arrival of the Chief Mate

The sun had scarcely shown his face over the crest of the hills when *Teddy* left Corme, all her canvas set to a beautiful northerly breeze. It was the first really fine day we had had at sea since the day when we left Norway, and I was naturally keen to make the most of it. The sea was deep blue, and *Teddy* slipped through the water with an easy sway, reeling off her eight knots without effort.

The spray that rose from her bows glittered in the morning sun; she reminded me of a young maiden playfully throwing into the air a necklace of sparkling diamonds, adorning her neck.

Oh, undoubtedly a beautiful day! But its name was Friday. And, as my wife had asserted, pretending the superstition of an old salt, it was tempting Providence to start on a Friday.

These apprehensions seemed justified. About seven o'clock that same night we broke our main boom. We were then some thirty-five miles south-south-west of Finisterre heading for Lisbon. It was all my own fault, fortunately. It is so easy to forgive oneself.

My wife and I were sitting in the cockpit; I did the steering and talking. Just then I had been discoursing widely on the feeling of security—one might almost say immunity—it gave me to know that the rigging was in such good order, that nothing could carry away. Hadn't I, myself, overhauled every bit of the gear since we left Corme? I grew quite enthusiastic about this point and forgot to watch the tiller.

And then she jibed.

The heavy mainsail came over with a bang and the main boom snapped off like a carrot!

As I have no means to repair the damage we were forced to put into Vigo.

It was a good thing that we had not come farther south. As it was, we could still head up to windward of Cies Island. Coastal and harbour charts were yet unknown luxuries to us, but we groped our way into the bay and, about midnight, found ourselves becalmed in the harbour.

Winged, as we were, we had some trouble in keeping out of the way of numerous fishing boats, who came steaming down the harbour at great speed, apparently, without ever looking ahead. Repeatedly I had to resort to the foghorn to call their attention to our presence.

At daybreak a light breeze sprang up and carried us to an anchorage close to the walls, where we remained for five days, during which time a carpenter converted into a main boom our former jib boom, which in any case I had found too heavy for one man to handle.

In Vigo we shipped a new member of the crew, a puppy. She had been born on a Norwegian steamer, and the chief, who gave her to me, claimed that she was a 'French police dog.' Her origin caused me no worry. Whatever her pedigree, she suited me. She is good-looking, very much like an Alsatian, but smaller, sweet tempered and obedient, but a fierce watchdog and withal the most intelligent canine friend I ever had. She took to me from the beginning; wherever I went she wanted to go.

I called her 'Spare Provisions,' which is still her official name, though as this title seemed to dishearten her, perhaps bringing to her mind unpleasant apprehensions, I usually addressed her as *Teddy*. This had the added advantage that if anyone hailed the boat when we were at anchor, the dog would immediately hear it and answer.

Although a proper sea dog, born and bred, Spare Provisions was very seasick on her first voyage in the *Teddy*. I had to carry her with me from the cabin to the cockpit, from the cockpit to the cabin. She seemed to be less miserable in my company; with her head resting on my sea boot she lay still, content to look at me. At times she would even attempt to wag her tail to show her gratitude.

We had a rough passage from Vigo to Lisbon, taking ten days to cover a distance of 260 miles. For those ten days the dog ate nothing, but when at last we arrived her appetite turned and she soon made up for her long fasting. She ate everything she could get hold of, including Vaseline, coffee beans, brown pepper, tobacco and a bank note equivalent to twenty pounds of fish. However, she deserved it all, and more.

We could not have had a better guardian for our boat. For seven weeks we moored alongside the quay at Lisbon. My wife and I went ashore every day, leaving the boat in her charge. She never allowed anybody to come on board during our absence, and it is certainly due to her that nothing was missing when we went to sea again.

Until we arrived at Lisbon my navigation facilities had improved but slightly. Here, however, I managed to buy a second-hand sextant. It was not too good, to be sure, but neither was I much of a navigator, and in any case, on a small boat, the conditions would rarely would be in favour of taking accurate sights.

Nevertheless that sextant gave me a feeling of importance, of superiority; I had become a proper captain, a navigator supposed to be capable of solving the great secret of navigation. My superiority, I fear, was however very much akin to that of a native sorcerer and not much better founded.

The weather remained cloudy all the way to Madeira, our next place of call; my dead reckoning took us there right enough. With the island in vision, the sun at last showed his face.

Here was my chance. I tried out my sextant and was rewarded by the startling geographical discovery that the island was one hundred miles further north than shown on the chart! Had it then shifted? However, on the next occasion that I consulted the instrument, when leaving the Canary Islands, they were revealed to be one hundred miles too far south! Thus, striking an average, we might consider our instrument about right.

We were becalmed off Madeira for two days—often I had to take to the sweep to assist an occasional catspaw to keep *Teddy* off the rocks—and then, towards night on the ninth day from our departure from Lisbon, a fresh southerly breeze sprang up.

Tired of drifting about within ten miles of Funchal, I decide to make port straight away without waiting for daylight. By this time I had grown quite accustomed to entering unknown ports at night: Vigo, Lisbon, Funchal, what did it matter when I had a breeze and a BOAT!

On this occasion we passed through an exciting time before *Teddy* swung round to her anchor in a tiny space of open water in the crowded little bay behind the ancient fortress.

Approaching, I could not until the very last minute make the anchor ready. Otherwise she would have rolled the chain overboard. In addition the brilliant lights along the waterfront blinded me in such a way that I could not have told whether we were three miles or three hundred yards off the shore. Next time, of course, it will be easy.

We remained at Funchal for ten glorious days. The harbour is poor, but the island is beautiful.

Leaving Madeira we narrowly escaped a hurricane which two days later devastated Funchal, breaking down the mole, swamping the lower portions of the town, stranding ships and drowning many people. We succeeded in keeping clear of the cyclonic centre by running away to the eastward until, at last, abreast of the little group of islands named Great Salvage, we ran into north-easterly winds and fine weather. We had, at last, left winter behind.

It is difficult to imagine the feeling of satisfaction I had that night as we were sailing easily along over a silvery sea, with a great moon rising over the horizon, whence the breeze came, and above us the stars glittering like a thousand bright jewels in a magnificent diadem. A feeling of confidence possessed me that all my troubles were over.

The soft wind, vaguely scented, drove the boat over the long swell with a pleasant gliding motion. From time to time the sails would come aback, when the ocean, as with a deep breath, lifted *Teddy* high on her bosom. This was, indeed, 'a *belle aventure*'.

At daybreak we sighted high above the clouds the regular cone of Pico de Teyde. It was then about 100 miles distant. The summit of this peak is more than 12,000 feet high, and it has been said that from its top, on a clear autumn day, one can see the island of Madeira 280 miles distant.

We arrived at Santa Cruz de Tenerife in the middle of the night, drifting into the harbour with slack sails.

We had spent more than a week in covering a crow-flight distance of 260 miles.

In the sunny port of Santa Cruz we remained for more than a month enjoying the glorious weather, the bathing, the beautiful surroundings, and the friendly attentions of our countrymen ashore and on board ships. We were taken by our Vice-Consul for pleasant drives around the island and also participated in the gaieties of the carnival season. The authorities at Tenerife and also the Royal Yacht Club treated us with exquisite courtesy.

From Santa Cruz we sailed to Las Palmas on the island of Gran Canaria. There we lay for a space of four months, for my son claims this town as his birthplace.

Situated in about twenty-eight degrees northern latitude in an ever-smiling blue sea the Canary Islands enjoy an almost perfect climate. As semi-tropical parts go, they are not very fertile; the country is rocky and rain is scarce. Santa Cruz had been practically without rain for three years when we were there. Nevertheless the islands supply a considerable portion of the bananas consumed in Great Britain and France. When irrigated, the soil is capable of intensive cultivation, yielding, I believe, as much as a bunch of bananas annually to the square yard. The irrigation problem is the all-important question of the Canaries, and its solution would surely render these islands immensely productive.

Half a century ago the cultivation of cochineal brought great prosperity to the islands. Cochineal is a bright red dye made from the dried parasite living on a certain kind of cactus, which grows profusely on every kind of soil in the Canaries. Before the discovery of aniline dyes, cochineal rendered the chief source of purple red for a variety of purposes, and the Canary Islands had acquired a sort of world monopoly for its supply. Nowhere would those parasites thrive as in these islands.

The introduction of chemical dyes, however, ruined the cochineal industry and caused it to be abandoned, until in recent years the liberal use of the lipstick by the fair sex has brought about a revival of the industry, although as yet on a smaller scale.

It is remarkable that the Canary Islands, the eastern-most of which are situated almost within sight of the African continent, escaped discovery until the fifteenth century. When Columbus called at the islands on his first voyage across the Atlantic, they had only recently been fully conquered.

The former inhabitants of the islands left no records and therefore their origin and the history of the islands before the Spanish conquest can only be guessed at. Some people would have it that those islanders were descendants of the old Norsemen, others point at obvious traces of their connection with ancient Egypt, while the romantic-minded claim that they were the last of the peoples who at one time inhabited the lost continent of Atlantis.

In these sunny islands my boy saw the light of day. He was born on 10 May 1929 in the hospitable house of my friend Antonio Curbelo, whose Christian name he bears. He was a solid little chap right from the start, weighing thirteen pounds on his arrival. When he was two weeks old, his mother took him for his first promenade in the park to the great admiration of the female population of Las Palmas. When three weeks old, he joined his ship. But when this heir to all our belongings—consisting, it is true, mostly of castles in the air—had reached the age of six weeks, we again weighed anchor and set forth across the wide ocean on Tony's first cruise.

Challenging the Atlantic
Provisions ~ Trade Winds and Spare Provisions

Considerable experience is demanded in making up a suitable larder for a small craft bound on a long voyage, especially if the limitations of the purse are of first importance. Climate, humidity, ventilation, cooking facilities, storage capacity, and even the movement of the boat have all to be taken into account. However, ten months on the boat had taught us a thing or two, so that when we left Las Palmas our stock of provisions showed a fair amount of variation.

Though we could not bake bread on board, we did not miss it much, as we had four kinds of biscuits.

Salt meat, pork, peas and beans we did not carry, chiefly because of the long soaking these foods require before cooking. On a small boat at sea it is difficult to find a place when a bucket full of water will not upset. Too often I had used bad language slipping on peas which should have been in a soaking pail. Tinned corned beef and canned soups were easier to prepare and served us for a fair variety of tasty meals.

We carried enough potatoes and onions for six weeks, and as much fruit and vegetable as we thought we could manage to eat before they spoiled. They lasted us for two weeks.

The chief items of our larder were:

30 small tins of corned beef
20lb of salt codfish
20lb of stockfish
15 (2lb) tins of cod's roe

15 (2lb) tins of fish balls
10 (1lb) tins of salmon
40 tins of soup, mainly pea
3 doz eggs, preserved in dry salt
25lb of margarine
1 tin of salad oil
20lb of rice
50lb of gofio
30lb of sugar
2lb of tea
10lb of green coffee beans
50 tins of condensed milk
20 (2lb) tins of green peas
4lb each of dried apples, prunes, and raisins
2 to 3lbs each of flour, corn starch, tapioca
100lb of potatoes
20lb of onions

—plus a few tins of sardines, paté de foie gras, California fruit, syrup, a few jars of jam and pickles, further carrots, cabbages, tomatoes, bananas and oranges, vinegar, spices, etc.

Among the items of our larder there is one, at least, which deserves special mention—gofio. This is a kind of flour ground from hot roasted whole wheat or maize.

This way of preparing corn has been handed down from the Guanches, the noble and warlike people who lived in the Canary Islands at the time of the Spanish conquest. The Guanches, as a distinct race, have disappeared, having been assimilated by their conquerors to make a remarkably fine type of people. Gofio, however, is as much of a necessity to the Canary islanders of to-day as it was to those of five hundred years ago.

Mixed with a little water, milk, tea or coffee, it forms a hard dough which is eaten with every meal. Its uses are manifold and its taste excellent. Naturally, it is both nourishing and wholesome.

The people in Las Palmas procure their gofio daily, hot from the mills. We had ours packed in biscuit tins, which were soldered before the contents had had time to cool down. It kept perfectly and formed a most valuable part of our daily fare for months. We regretted our inability to procure it anywhere else.

At the dawn of the fourth day after leaving Las Palmas, I sighted the Pico de Teyde just above the horizon. Mustering all the fragments of my shattered mathematical knowledge I endeavoured to figure out the distance. When my efforts failed to lead to any acceptable conclusion, I called it 150 miles and was content to leave it at that.

So far we had encountered only light and variable winds but on that day we ran into the trades. Rapidly increasing, the breeze soon converted a modest wake into a broad band of foam. From noon until six o'clock we covered a distance of fifty miles, by which time, however, the sea had become so rough and the helm so hard to manage that I chose to lower the mainsail rather than to risk a jibe. Thereafter for more than one month the mainsail remained furled, the wind keeping practically dead astern all this time. With the headsails spread goosewing-like, one on either side, we made very satisfactory headway, some days averaging more than seven knots. Furthermore, the motion of the boat was easier, and she did not require such careful steering.

As we sailed the immense wastes of the Atlantic the force of the trade wind kept steadily increasing. The blue of the sky above became covered with a haze that grew denser day by day and made the sun look down upon us with growing indifference.

The clouds, which had at the outset travelled peacefully enough, clear patches of white and gold high up in the endless blue heavens, kept slowly sinking. Day by day they came nearer to the sea level, gaining speed, but losing form and light as they did so until finally they seemed to detach themselves, grey and woolly, from the dense mist close astern, hurry by, and again fade away in the dark grey nothingness that loomed everlastingly ahead. They were like formless spirits hastening towards Nirvana.

Towering seas shot by, now laden and menacing, now roaring angrily, as they swept past with thundering masses of seething foam.

Everything was hurrying towards the same goal with ever increasing velocity. There was something despairingly inevitable in this apparent rush of forces to destruction. It was like living in real life a sinister tale by Edgar Allan Poe.

In this world of restless motion *Teddy* tumbled drunkenly onward to her unknown fate. I could not help it, but responsibility lay heavily upon me at that period. No use repeating to myself my frequently expressed sentiment, that we must all die sometime. There was our boy; he had not chosen for himself; he had a right to live. Thus did I ruminate, although only a few days before I had scoffed at those meddlesome persons who declared that it would be nothing short of murder to take the child along with us on the cruise.

Gradually, however, I regained my confidence. After all, there are dangers ashore also. In fact, all through my previous years of seafaring I had found that the dangers of the sea were mostly ashore.

It cannot be denied that life at sea is a healthy one. Left alone, the boy throve; there was no one to worry him, to awaken maternal anxieties, or to poke needless pills into his small and smoothly operating interior.

Granted that I was a poor navigator with poor navigation facilities at my disposal, at least I knew something about seamanship. I had served eight years in square-rigged ships in my younger days. I knew how to handle my boat, and my faith in her knew no bounds.

On the tenth day out the wind reached its greatest force, and thereafter the weather gave us little reason for complaint.

The days passed on in full contentment and tranquillity. We sailed only in the daytime and hove to at night. Backing the staysail and hoisting a trysail abaft the mast, our little ship practically stayed on the same spot until we resumed our course.

When crossing the ocean, we did not carry side lights. Confident in the knowledge that steamers hardly ever come to those parts of the Atlantic which we were traversing, we went to sleep, as good people should do at night, quite undisturbed by any anxiety about being run down.

The whole ship was dark save for an inoffensive riding light, which I used to hang in the windward rigging and which made noble efforts to shine for the double purpose of attracting flying fish and of indicating our existence to any other ships that might have lost their way. However, it generally went out before midnight, and I never relit it.

Tony used to wake up once every night, when he would voice his resentment against the rolling of the ship and the conditions in general, but these complaints were always cut short with the appearance of his bottle. Like father, like son! was my wife's comment.

Coming on deck in the morning we occasionally found some very obliging flying-fish waiting to be fried for breakfast.

Our days were fully occupied. Preparing meals, washing dishes, mending sails and gear, washing clothes, bathing, fishing, kept us so busy that we had little leisure for reading.

We took shifts at the tiller, my wife and I. The baby generally slept in his little cot in the cockpit. Even in his sleep he kept a firm hold of the sides of his cot to steady his cherubic body against the rolling of the boat. I think it developed his muscles. He was becoming a regular sailor. One day, in perfectly fine weather, a sea broke in and filled the cockpit, where the child in his cot, much like Moses, was set seafaring on his own. He laughed with delight when his mother rescued him!

By this time, however, we had come well south; the end of July was approaching; scorching sunshine made a bathe a very pleasant thing. It was a luxury which my wife and I enjoyed perhaps four or five times a day. For weeks we wore nothing but bathing suits.

Once or twice I took my bath outside the boat, hanging on to a line, but when sharks began to keep us company, sometimes following in our wake for several days, we were content to take shower baths on deck, hauling up water from over-side with a bucket.

The boy had his daily bath in fresh water; it was never heated, he had been accustomed to bathing in cold water since he was born.

Spare Provisions proved a good shipmate, both amusing and sensible. She would often play around the deck with ropes, coil them all around herself, get entangled in them, then growl and shake them with her teeth and make much noise.

That was her way of taking exercise after a good feed, as for instance when we had caught a fair-sized fish or when, for some other reason our dinner had been specially abundant. When it happened that she had been reduced to little or nothing but biscuits and water, she would lie down under the table and avoid every unnecessary waste of energy.

As long as we had stockfish on board, such periods of sparse living for the dog did not occur, but later on, these periods were not infrequent. Once, when Spare Provisions had been on a diet of biscuits and water for several days I took pity on her and, in her presence, opened a tin of fish balls, the contents of which she devoured in record time. When, on the following day, she found that her fare had again been reduced to only biscuits and water, she went into the forepeak and fetched out, one after the other, all the empty tins, which I had put away for use as paint pots. She brought them to me, asking me in plain dog language to produce fish balls from out of those tins. I told her that I was sorry, but that such a trick was beyond even my powers. She said, 'I am sorry, too!' And then proceeded to eat her biscuits.

She is both very affectionate and very jealous. She seems to resent my wife speaking to me and when something amusing invokes our laughter, she jumps up and expresses her disapproval by loud barking. Towards Tony she has adopted an attitude of suspicious and cool reserve. And yet, when told to look after him, she will do her best and remain rooted to his side.

At sea we generally took our meals on deck, standing in the cockpit and balancing our plates, while the bights of sheet ropes served as fiddles to prevent pots and pans from tumbling about.

Spare Provisions would remain below until she was called or until the noise of dishes told her that our meal was over. Then she would appear in a flying leap, like a canine comet.

As I did not want the dog to eat from the dishes we were using, I provided her with her own special enamelled plate below, which I taught her to bring to me. It was only necessary to explain this once; thereafter she carried out the action immediately I gave the word. It would have been easy for her to acquire all manner of tricks but why should I worry the dog with such foolishness?

She has proved an excellent retriever. That, however, is due to her nature, not to teaching. Her love of retrieving is only equalled by her joy in swimming—two gifts which have proved highly useful when articles have gone overboard.

Whenever opportunity offers she takes to the water, but she is sensible enough to keep well out of it when we are speeding along or when there are sharks about. I discovered her love of swimming and retrieving when, one day at Funchal, I caught Spare Provisions enjoying herself by throwing overboard the dinghy's baler, and then jumping into the sea to fetch it out again.

Every night at Las Palmas I encouraged the dog to go for a swim. We were at anchor and she could not climb on board without assistance, as *Teddy*'s lowest freeboard is thirty inches. One night, while Spare Provisions was having her bath outside, I received visitors and took them down into the cabin, where we sat yarning for many hours, till well past midnight. When I came on deck again to see my guests off, I found that my faithful and neglected dog was still swimming round and round the boat. Her master had forgotten all about her. She betrayed no sign of exhaustion, however, and after she had been pulled on board and had vigorously shaken herself, she pricked up her ears and stood expectantly looking at me as if asking: Well, what next?

Had she really been tired, she would surely have had sense enough to swim ashore for a rest. The steps were only some 200 yards away.

At Las Palmas we would sometimes take the dog along when we went ashore. Spare Provisions loved to go wherever I went, even if it meant leaving the yacht. While we were dressing my wife and I would discuss whether or not to include her in the trip. That dog understood every word. If the decision expressed was favourable to her coming, she would leap on deck, dance around the companion and bark down the hatch as if saying: Hurry up, you slow people; come on, let's be off!

If, on the other hand, we had decided that she should stay on board, she would then crawl dejectedly under the table, utter no sound whatever and hardly stir even when I bent down to pat her good-bye. There she would remain, grieving, until I came back again, unless it chanced that she was stirred up by some stranger trying to come on board.

On one occasion, we happened to be away thirty-six hours. I had left the dog plenty of biscuits and fresh water. When I returned, I found that in her mood of misery she had touched neither, whereas immediately the welcoming ovations were over, she went for it with a will, finishing off the food and lapping up half a gallon of water in the wink of an eye. Sometimes even throwing the biscuits high into the air in her excess of joy.

A queer sort of dog indeed is Spare Provisions.

7

Life on the Ocean Wave ~ Problems of Navigation
A Meeting at Sea ~ An Exciting Run and a Landfall

Some wonderful days we spent in those north-east trades.

The sea was blue like a dream. I do not know how I could otherwise describe this transparent beautified reflection of the immense heaven above, and the heaven in itself was a revelation, a dark blue infinity, where the trade wind, like an invisible shepherd, drove his shining sheep in herds of everlasting procession.

As *Teddy* cut her easy furrow through the waters, she would scare up schools of flying fish. Sometimes they were chased by bonitos, dorados or tunas. In the clear water we could follow their progress far away.

Now and again we would see huge fish deep down below *Teddy*'s bottom, twenty fathoms or more. When we were not travelling we could see much further down. For several minutes I have been able to follow the flashes of an empty tin I had thrown overboard. However, since the ocean in these parts is nearly four miles deep, it would yet have a good many hours to travel after I had lost sight of it.

The great event of the day occurred when I had taken my noon altitude and calculated the ship's position. What a thrill it gave us! To be sure, the high seas and the rolling of our craft certainly interfered with the accuracy of my observations. However, by taking noon latitudes every day, unless the sun was hidden behind clouds, I obtained a continuity in my sights, which tended to stabilise and improve my nautical guessings. In order to check my latitudes I used once in a while to measure the angle of the polar star. It is true, I did not know which corrections to apply, but fortunately this star is so conveniently placed that the angle of elevation in itself gives the latitude within a degree or so.

When according to my dead reckoning we were about 46 degrees west and 13 degrees 40 minutes north latitude, we sighted a steamer. I was not surprised to see her. In fact, I had been looking for steamers. We were in the track between New York and all the principal ports on the east coast of South America.

The ship may have been some seven or eight miles away and was about to pass by, when all of a sudden she altered her course and came towards us.

She proved to be the Brazilian ship *Alegrete*, probably on her way to the river Plate or some other South American port. When she was near enough I signalled to her, asking for my longitude. Her reply gave me great satisfaction; her longitude differed only 31 minutes from mine. I readily admit that it was sheer luck, and neither this nor any later case of well-nigh astonishing accuracy in my nautical findings has changed the attitude of distrust I maintain towards my navigation. To be safe at sea, I think one should regard one's nautical calculations with a great deal of scepticism. This applies more particularly in my case owing to my lack of proper instruments.

After meeting the Brazilian I shaped a course more to the southward. I had two reason for doing so; I wanted to avoid the waters east of the Windward Islands, whence most of the West Indian hurricanes originate in August, and I wanted to come near the steamship tracks with a view to correcting my position once more before approaching land.

On August 3rd we began again to look for ships. It was a week after meeting the Brazilian, and according to my calculations we should be in latitude 11 degrees north and longitude 55 degrees west, in waters where I expected to find a good deal of traffic. All the shipping from the Mexican Gulf and the Panama Canal to the eastern ports of South America, to the Cape of Good Hope and beyond passes this way.

However, the day went by without bearing into sight a single ship. The weather was fine and the visibility good.

The next morning I hoisted the mainsail. It had been furled since

July 3rd. At noon I figured out the ship's position to be 10 degrees 44 minutes north and 56 degrees 9 minutes west. I climbed the rigging a dozen times but saw not a ship.

On August 5th my reckoning put us 90 minutes further west in the same latitude. We were in a part of the sea where the northern and southern Equatorial currents unite and set strongly to the westward, particularly in the prevailing rainy season in South America, when the flooded Amazonas with its enormous masses of water materially accelerates the coastal current. I had been told that its velocity often exceeded 3 miles per hour at that time of the year. Under the circumstances we might easily be a hundred or even two hundred miles west of our calculated position.

But even if, for these reasons, the accuracy of my longitude was very doubtful, yet we ought surely to have encountered some ships, if my latitude was correct, which, by this time, I began to doubt.

The situation threatened to become exciting.

Where were we?

One thing I knew: we were not very far off some land or other, as was testified by the presence of large flocks of seabirds.

That night we lowered the mainsail, before we hove to.

August 6th broke cloudy and with light winds; the visibility was still good, but not a streak of smoke showed above the horizon, not a sail.

I did not get any observations on that day; our noon position by dead reckoning worked out to be 10 degrees 24 minutes north and 58 degrees 23 minutes west.

Neither did August 7th give me an opportunity for a shot at the sun. Our calculated position at noon was 10 degrees 25 minutes north and 59 degrees west. To be prepared for surprises, which might crop up at any time, I hoisted the mainsail again. I thought it advisable to have my boat in more complete control so that she would readily handle.

Understand our situation: 'Sailing Directions' emphasizes that no ship should attempt to make a landfall in these parts without first having made sure of her accurate position. This is a most reasonable precaution. Rocks awash and far off-shore banks in connection with a sweeping tide and a heavy sea make the approach extremely dangerous even for well navigated steamers. They must know where they are 'before' they make a landfall, because those dangers are beyond sight of land. As for us, for all my dead reckoning we might be anywhere within a circle of some 500 miles diameter.

All this, of course, I knew previously, but I had relied on meeting a steamer to correct my position, and now we had crossed all steamship tracks without even meeting one.

August 8th. Overcast. No sights. Visibility good but no ships to be seen. Calculated position 10 degrees 18 minutes north and 59 degrees 54 minutes west.

In the afternoon the wind freshened to a moderate gale that gave *Teddy* a speed of 8 to 9 knots. This brought the situation to a crisis.

To heave to and commit ourselves to the mercy of the sweeping current did not appeal to me, because such a step would have rendered us helpless in the event of sudden dangers cropping up. It was better to keep *Teddy* under plenty of canvas and moving.

But I had perforce to decide on a course with nothing to trust to but my own resources.

Considering that we had met no ships, I chose to build on the assumption that our latitude—the only feature in my navigation that was based on something more than guesswork and intuition—was wrong, and that we were further north than calculated. I therefore changed our course to south-west.

South-west certainly was a dangerous course to steer in the case of my nautical findings being right after all. But was that likely?

Furthermore, at the time I could not see any alternative which did not involve just as much risk or more. I had to 'give it a go.'

To approach a dangerous and unknown coast at that speed was certainly a severe test upon our nerves—a particularly eerie experience after the darkness of a moonless and cloudy night had settled upon the narrow circle of tumbling waters within the limits of our horizon.

I was prepared to see the breakers rise around us at any moment, and the situation became ever more exciting. The sea was heavy and awkward; it was evident that the current was swift. The night seemed crowded with dark apprehensions. As I stood at the helm trying to pierce with aching eyes the wall of blackness ahead, a huge comber would suddenly arise in our way, break with a mighty roar and a tumble of phosphorescent foam that would shine greenish white through the darkness. Then, tense with excitement I would stand there gripping the tiller tightly, ready to grab the sheets and bring the boat to instantaneously.

My wife stood by my side praying into the night. She knew what she had to do. I had instructed her. Tony lay sleeping peacefully in the twilight of the darkened cabin. The heaving of the ship or the outcome of our hazards left the wee chap unperturbed. He even smiled in his sleep. Beatific smiles that revealed his dimples.

And all this time we were actually leaving the safety of the steamer tracks behind.

Of course, we did not know. But then, just as we were racing over the hill-top of a wave, my wife, while chancing to look astern, gripped my arm.

'I think I saw a light on the port quarter,' she said excitedly. I looked, but could see nothing. We both looked. Nothing but blackness and the phosphorescence of the breakers! Yet a queer uneasiness would not allow me to dismiss the question. I handed the tiller to my wife and climbed the rigging.

Ho! There was a light, sure enough.

I climber higher until, standing on the swaying gaff I made out, whenever we were on top of a wave, two lights close together, one a little higher than the other, the masthead lights of a steamer, which I judged to be steering a north-westerly course. It struck me that she must be heading for the straits between Trinidad and Tobago, coming from Cape San Roque.

Acting on the impulse I clambered down, brought *Teddy* about, and let her come away on the other board until she headed north-west with the new wind well abaft her beam.

Then I went below and consulted my chart. It seemed as if my dead reckoning had been correct, after all.

Working on this, I shaped a course for the Straits, north-west a half north. Three hours later my confidence met a welcome support in the shape of a big liner steering a parallel course.

At 2 a.m. we hove to: if my calculations were right, we ought to be just off the straits. That night we burnt our side lights for the first time since leaving Las Palmas.

At daybreak, when the weather cleared up after a heavy squall, my wife suddenly put her arms around my neck and gave me a hug. The first rays of the sun had conjured forth the outlines of an island, high wooded mountains:

TOBAGO!

That was Friday, August 9th.

What a wonderful piece of luck it was that my wife chanced to espy that light. Five minutes later it would have been invisible even from the masthead, nor would any other lights have come our way. Before morning we should have been in the breakers on the banks off the river Orinoco. And yet, was it really the merest chance that prevented us from continuing our mad race to destruction?

8

Sailing the Caribbean ~ The Dutch West Indies
My First Lecture ~ A Speedy Passage and Rats

As we sailed through the straits, we generally had a clear view of some part of Tobago between the squalls, whereas Trinidad kept permanently hidden behind clouds. Grey fragments of vague outlines would reveal themselves on rare occasions, but such never sufficed to give us a bearing. The weather was most unsettled, bright sunshine alternating with fierce squalls and frequent heavy downpours. All around the water was of a dirty brown colour; on its surface floated twigs, leaves, green plants and even large trees, uprooted and carried away from the forests on its banks by the swollen waters of the mighty Orinoco. In the rainy season this river discolours the sea over an area of hundreds of miles.

The current in that passage, everywhere it formed seething rips and whirlpools, sometimes turning the head of my boat many points off her course.

Once past the Bocas de Dracos—the Dragon's Mouth—which forms the narrow and tricky entrance into the Gulf of Paria, between the mainland and Trinidad, we came into still waters and continuing our course along the Venezuelan coat we met only light weather.

The forests of Venezuela, rising like an amphitheatre towards the high mountains in the interior, were continually covered with grey mist and clouds. Only on rare occasions would the sun paint gay green spots on some grassy hill along the coastline. More frequently the coast and everything beyond it would be obscured from view by dense thunder-clouds that crowded growlingly along the border of the sea as if begrudging us the fine weather. Surely it was remarkable, for there at a distance of only five miles from the shore we were sailing leisurely along under an almost perpetually blue sky. The dirty weather prevailing on the continent formed one of the reasons, indeed, why we did not call at Cumana. As it happened, it was lucky that we did not, for just at that time a more than usually energetic revolution in Venezuela rendered conditions rather lively there, Cumana being one of the centres against which the rebel general Urbino directed his most ardent activities.

Following the coast we left the islands of Margarita and Tortuga to port, but otherwise we went inside the chain of islands, cays and shallows, which stretches for some 500 miles from Testigo to Aruba, at an average distance of about 80 miles from the mainland.

Crowds of pelicans and other birds kept us company. Porpoises were playing around everywhere. Schools of fine big flying fish were permanently about, but not a single fish came on board. They must have read on our greedy faces that our interest in them was inspired mainly by predatory motives, and in that interpretation they would have judged aright. Our ship's fare was beginning to grow somewhat monotonous. We were longing for a change.

At length, after forty-eight days of sailing, we arrived at Curaçao, the principal island in the Dutch West Indies.

This island, formerly of very little use to anybody except, perhaps, in olden days to the buccaneers, has lately become one of the world's largest shipping centres, on account of its oil refineries.

Apart from its natural harbours the island possesses few attractions, being almost barren save for divi-divi trees and enormous cacti. As a general rule the trade wind blows strongly all the year round, but we experienced a two weeks' spell of dead calm when the average temperature day and night was 94 degrees Fahrenheit, while on one occasion I even read 97 degrees at 7.30 in the evening. The atmosphere was suffocating.

Yet, in the solitude of the little bay where we were anchored, we enjoyed nights of rare beauty, enchanting nights, filled with the strange odours of the island. We might be sitting on deck for hours without uttering a word, just watching the moonlight playing on the drowsy waters of that still lagoon and listening to the weird song of the crickets forming an eternal concert which, rising and falling as at the command of an efficient and temperamental bandmaster, sounded like the ever-changing whistle of a gale in a hundred telephone wires. It was in that little bay we spent the best part of our stay in Curaçao. Alongside the wharf, where the harbourmaster had at first most courteously berthed *Teddy*, it had been impossible to remain on account of the overwhelming interest shown us by the coloured population of Wilhelmstad.

At the anchorage we were rarely disturbed except by our friends or by countrymen who came to pay us a call or to fetch us for a quiet yarn aboard their ships. Such pleasant interludes were frequent, thanks partly to the great number of Norwegian ships calling at the port, and partly to the hospitality of the Norwegian Consul and his mother, Mrs Maduro, the grand old lady of Curaçao. Thus we had an excellent time generally in spite of the thermometer.

Meantime the intense heat and the sunshine nearly proved our undoing. As we discovered after our departure, the planks in the hull of the *Teddy* had shrunk so as to cause serious leakage above the waterline. We struck boisterous weather outside with the result that our boat drew water faster than I could pump it out. Fortunately the island of Aruba was only seventy miles distant, and we managed to make port in San Nicolaas Bay with two feet of water in her hold.

I decided that it was necessary to get *Teddy* out of the water and give her an overhaul. That would cost money and I had none.

Because guineas are of no value at sea I had, as usual, exhausted my monetary resources before leaving port at Curaçao, not finding much to write about on the island of Aruba, which is only a smaller copy of Curaçao. I tortured my brain to find a solution to my difficulties. At last a bright idea came to me. Eureka! A lecture.

True I had never before made a speech in English and I naturally entertained some misgivings as to the outcome, but knowing that tomatoes were scarce on the island I thought that I would brave it. Nothing venture, nothing have.

The large oil refineries in San Nicolaas Bay, employing about a thousand white men, would certainly yield victims to my enterprise.

Therefore, putting on my best and only shore suit of heavy blue serge, I went to call on the Managing Director. Very slowly did I walk in that suit. For the heat was overpowering and it was vitally important that the high chief should not receive a flustered or perspiring caller. Provided I appeared cool and composed, he might not notice the arctic nature of my attire.

His reception was of the kindliest and immediately I had outlined my plan, he expressed his entire approval, promising every support that I could possibly desire.

Naturally it proved a success.

Although I make no pretentions to illuminating, like a newly arisen star, the world of oratory, the spirit of cordiality that prevailed in the assembly made talking easy, and my shortcomings as a public speaker passed unnoticed.

Those people of the Pan-American Petroleum Co. at Aruba were exceedingly decent to us.

While at Aruba, Spare Provisions had a litter of pups of unknown ancestry. They were, apparently, a combination of all the famous breeds of dogs in the world; the outcome was so varied, indeed, that each of their four legs seemed to be inherited from a separate breed of ancestors, differing accordingly in shape and length. Yet the pups were in great demand with the coloured population of San Nicolaas. If I had been a businessman, I should have been able to sell them at ten to twenty dollars apiece, and that is certainly what I ought to have done by way of compensation. But how was I to suspect, when I gave them away, that every flea-ridden individual in that mongrel company was destined to be named after my boat.

The necessary repairs effected, we sailed the 800 miles to Panama in four and a half days, so far the best long run we had had.

Until we came within twenty miles of the isthmus all we carried was a square sail, a bit of heavy canvas that had at one time been part of an engine-skylight cover on a Norwegian steamer. The total area was hardly 150 square feet. The sail and the heavy sweep, which I used for a yard, were continually under such a heavy strain that I might have expected the whole outfit to carry away at any minute. I feel convinced that it would have done so too, if I had allowed the sail to shake but once. I would have surely taken it down, if that operation had not been so awkward. It would have meant leaving the boat temporarily without canvas in the heavy breaking sea, and it would have taken hours of labour under the most trying conditions to set things right.

Therefore I preferred to chance it and was duly rewarded, although the voyage undoubtedly proved a rough and unpleasant affair.

While at Aruba rats had invaded the ship. This invasion might, of course, be regarded as a reassuring evidence of our boat's seaworthiness; if it is true that rats will leave a ship destined to founder, they would surely not embark in a vessel, when there could be any doubts regarding her ability to keep afloat. In a way, one might consider it a certificate of seaworthiness. Nevertheless, we did not enjoy their company. They were insolent and familiar, sharing our bunks and our meals uninvited, and even playing about on the cabin table, while six or seven people were sitting close around it.

Therefore, whether they constituted a certificate of seaworthiness or not, from the day they came and until finally they were treacherously potassium-cyanided to death at Cristobal, a bitter and incessant war was maintained against the invaders.

9

Through the Panama Canal
The Market Value of our Chief Mate ~ Golden Prospects

We arrived at Cristobal on November 24th, on the heels of the last squall of the rainy season. From then on and during our entire stay in the Canal Zone we revelled in glorious weather. The harbour authorities found a pleasant and convenient berth for our boat, just opposite the house once occupied by the French engineer, de Lesseps. A gangway plank connected us with the shore. Here we remained for seven weeks.

We lived on board, where I did my writing, whilst my wife and baby went promenading beneath the palms. I had begun writing a Norwegian book on our travels.

Whilst watching our chance to get through the canal, the days passed pleasantly enough in enjoyment of motoring, sight-seeing and courteous hospitality generally.

Having no engine and being hardly in a position to afford $100, which the towage through the canal would have cost, I was on the look-out for a small ship willing to take me in tow.

At last it offered itself in the shape of an American motor yacht, which gave us a rope and towed us some two or three miles, as far as the Gatun Locks, but there left us in the chambers. Meantime, having once entered the canal and the wind being fair, I saw no reason why I should not hoist my sails; which I did; and so it came about that we passed through the Panama Canal as far as Gamboa under our own canvas. We had no pilot on board.

It was a glorious night. The full moon showered silver over an enchanted world of sleeping jungle. There is no traffic on the canal at night. When day fades the Gatun Lake and its surroundings become an integral part of the neighbouring wilds.

It seemed as if my wife and I, with the baby wrapped in sleep below, were the sole human inhabitants in that silent vastness.

The wind was a mere vague breath, causing not a ripple on the polished surface of the dark waters. *Teddy* glided calmly on, as if by magic. No wake showed her way. The sails hung in huge folds, immoveable. Not a block creaked; not a rope shifted. At times we would pass so close to the shore that our boom end would scrape the outreaching branches and stir up a many-voiced protest from the sleeping inhabitants of the jungle. Alligators would splash noisily into the lake. The angry roar of a puma would call for silence. Then the voices would subside. Soon again silence brooded over everything. Beneath the dark branches swarmed myriads of fireflies.

We reached Gamboa. A train stood coughing at the station. A voice called out across the water, '*Teddy*! Ahoy!' We were expected.

We passed the remainder of the night at Gamboa and in the morning a launch towed us through the Gaillard Cut as far as Pedro Miguel. At Miraflores we had to tie up and wait for the north-bound ships to pass through the locks. When our turn came to enter the chamber, we had no motor-boat to assist us, but, hoisting our staysail, we sailed in. I was determined that my *Teddy* should not be accused of delaying the traffic.

In the afternoon we passed through the last lock and a launch sent by the port captain towed us to our anchorage off the Balboa Yacht Club's boathouse. The waters that washed around *Teddy*'s bows with the incoming tide were the waters of the Pacific.

For more than three months we remained at this anchorage, enjoying the hospitality of some charming people and befriended by all. Interesting excursions by boat and motor-car took me away from my work more often than they should have done, but when the rainy season was again drawing near I had to hurry my book to a conclusion.

Before we left, Tony had learned to walk. He was not too sure of his legs, perhaps, but at least he had the orthodox swaying gait of a sailor.

An efficient climber was Tony already. When he first climbed the seven steps of our nearly perpendicular companion ladder, only ten months had passed over his baby head.

From then on we had to keep him under the strictest surveillance. I made a gate at the bottom of the steps, so that he could not attempt this climb without our supervision, and also an ingenious contrivance of a strong canvas harness with a strap at the back to tie him up when on deck. We could not afford to lose him.

Tony seemed to be a source of anxiety to not a few people. A lady we had met some months before wrote and offered us a compensation of £600 if we would let her have Tony. The offer was made in all sincerity, because she and her husband had become fond of him and were anxious about his safety. They were kindly and genuine people of independent means but without children of their own. Realising as we did the motives operating, we could feel only gratitude for their solicitude. On the other hand the incident furnished too amusing an aspect to overlook. We could not help frequently jesting about this novel fashion of making a lucrative livelihood, and enlarging in imagination on the enormous prospects it offered for development.

Well, this generous offer we refused, but perhaps some day I may be sued for damages by an adult son for having disregarded his best interests by doing so.

To be sure, at this particular juncture, a paltry £600 offered little attraction. I had decided to go to Cocos Island and lift the treasure. Oh, of course there is a treasure. No romantic person would doubt that. Three million pounds in gold bars. It was just waiting to be collected. You have only to glance at the chart. Cocos, you will find, is only a tiny speck in the ocean, as big as the point of a needle, a delightfully limited area to hunt in for so much gold. In addition, the island is uninhabited, no one to dispute your possession. Oh, it was surely an easy programme, merely a sort of business journey to collect overdue money. I, being in a fashion a pirate myself, was naturally entitled to it.

It would come in handy. Those six tons of pig iron which we carried inside as ballast caused me considerable worry. In corroding, the iron deposited a thick layer of slime and gravel on the bottom, which often clogged up the bilges, a serious matter for a mariner. Now, gold does not rust. Yes, it would come in handy.

Besides, the thought of financial reinforcements was not altogether out of place. My ready cash in hand and otherwise had dwindled down to four cents. This was not due entirely to outrageous living. I had bought provisions for one month. I had bought ropes to replace my running gear which was worn and, after a fire on board, I had invested in a shining brass fire extinguisher.

Though to some it will seem hazardous to go to sea in a sailing boat stocked so inadequately with provisions, a moment's reflection will reveal that the blame is attachable to others than myself. Had prices been reduced by half, I should have laid in twice my present store—and so on proportionately to the extent of the reduction.

Tony at least was well provided for; we had powdered milk sufficient to last him for four months and, of no small importance, we had an ample supply of ship's biscuits.

Our stock of potatoes and vegetables was somewhat inadequate, potatoes to last for three weeks, perhaps, and vegetables for a few days only. Here was another reason for calling at Cocos Island. Judging by its name and the reports of numerous people who had never been there, the island abounds with coconuts, lemons, oranges and wild pigs. Really, there seemed to be little reason to feel gloomy about our larder.

Moreover, a friend of ours, an ardent fisherman, had supplied me from his own ample store with a wonderful selection of fishing gear, which gear I estimated the equivalent of many fine dinners.

And thus, after cleaning the boat's bottom and filling her water tanks, I picked up the anchor and with a light heart set sail to venture forth across the lonely Pacific, the vast expanse of South Sea, with its myriads of islands full of romance and mystery, and the underlying lure of buried treasure.

10

The Bay of Panama ~ Doldrums ~ And a Battle of the Elements

We left on April 17th, and the rainy season had set in.

The Bay of Panama is a tedious enough place for a sailing vessel to navigate at all times of the year, but more particularly is it so in the rainy season. The prevailing wind—if such a term as prevailing can be applied at all—is south-west. An entrance to the bay is generally easier to negotiate than an exit. However, the Bay of Panama and the waters west of it to Cocos Island and beyond are in reality an area of doldrums. A full description of the weather conditions of these regions would demand much time and patience from both writer and reader but, speaking in general, the rule is light airs and calms alternating with thunder storms and squalls from varying directions.

Around Cocos the unreliability of the conditions of navigation is further accentuated by unruly ocean currents, which change force and direction in a most unexpected manner.

It is, therefore, not in the least surprising that even the great seafarer, Captain Dampier, having set out to determine the exact position of Cocos, failed to find the island. After cruising about for six weeks he finally retraced his course in the firm conviction that Cocos Island was a myth.

Certainly, when Captain Dampier set out, he was ignorant of the exact position of the island, but as to that it was only approximately that I knew my own position. And yet the idea of missing Cocos never entered my head. I was accustomed to rely on my luck. It was only after I had witnessed the mysterious way in which this island sometimes plays hide and seek, wrapping itself up in a cloud or hiding behind squalls, so that one might easily sail by at close range without any idea of its presence, that I realised my luck in finding it so readily.

The Bay of Panama and the adjoining waters are alive with fish of many varieties. Every little while *Teddy* would be surrounded by dense schools of them. A few I speared, others I caught with a short line, but no opportunity came my way of trying my luck on big game fish. Before we rounded Cape Mala, our slow progress was not in favour of fishing with spinners, while afterwards matters of much greater moment occupied my attention.

This was rather unfortunate, however, seeing that the fish abounded in these waters and huge skates, some of them surely weighing several tons, were jumping about and disporting themselves continually. Sometimes they made me jump too, as they fell back with a tremendous resounding splash into the sea.

Since there was hardly any wind except in the squalls, I had to make the most of them, while they lasted. This meant running a certain risk, because one never knew what they might bring. The range of their possibilities lay between a gentle shower and a howling gale. Once or twice I lowered the mainsail before an approaching squall and rejoiced afterwards at having done so, but more frequently I was caught and had to make the best of conditions. This latter state of affairs generally occurred at night, when it was not so easy to define the nature of the squalls or even to see whence they were approaching. At times it was rather exciting.

But all these troubles pale into insignificance when I think of an electric storm we experienced during the night between April 28th and 29th. This storm was of such demonic violence, and accompanied by so many uncanny side phenomena, that the bare thought of it is even yet exceedingly unpleasant to me.

The night, moonless and densely clouded, had settled around us with pitchy darkness. One could not see a hand held before the eyes. Rain came down in torrents. Repeatedly it drowned the riding light, until I abandoned the attempt of relighting it. Only the binnacle lamp was burning, throwing its faint rays on the cockpit coaming, on a hand that was groping for a sheet rope or a shining black oilskin coat.

And then the tempest broke loose. Crackling lightning bore down from out of the greenish-brown poisonous-looking clouds, that crowded low above the phosphorescent masthead. In ever quicker succession it swished down into the sea to right and left, some of the flashes in the immediate vicinity of the quivering boat, while the incessant roar of thunder, deafening the ear, shattering the nerves, sounded like hell let loose, like the infernal gunfire of a million gigantic demons at war.

The wind kept veering round from one direction to the other, blowing sometimes with hurricane force. I stood at the tiller tensely watching and running before the gale in order to save the rigging.

Then there would be a lull, intervals between the bursts of lightning—intervals of utter darkness.

I could see absolutely nothing. My only object was to keep the wind well aft and watch out that she did not jibe. Whatever I did, I must not jibe!

Gradually I could discern the greenish glow of those phosphorescent spar ends, until new flashes of lightning would make every detail distinctly visible in a ghastly bright light. I can still see that weird intensified picture of the deck asplash with the downpour, which came in hissing gushes, of the shiny black rigging and of the straining grey canvas, streaming with driving rain.

Then the horrible outburst would recontinue, fiercer, apparently, and more fiendish than ever. At times the whole sky would be a dense cobweb of lightning, flooding every crack and corner with an abominable brightness, and then again, in the blackness that followed, we would be rushing through wide streaks of dully glowing water which stood out sharply, without surroundings, without background, like a flood of luminous milk in an empty space. What it was, I cannot tell. Perhaps the phosphorescence caused by billions of microscopic beings, isolated by meeting currents—perhaps only one more amongst many weird electric phenomena, which added to the indescribable horror of this night. More than once I almost expected the whole universe to explode.

According to my calculations, we should be close to Cocos Island. What if we ran into a reef? That would bring the end, extinction. But what could I do? I had no will and no power but to obey the elements, rush onward, somewhere, anywhere...

To oppose them would be fatal.

How utterly futile, however, it is to attempt to describe what one cannot even understand. What are the words? What are the epithets? Empty, surely, and without force or meaning compared to the horrors of that night, a night which gave me a more vivid and terrifying picture of inferno than my imagination could ever have created. It was like a weird prelude to the Day of Judgement. And when, at last, morning broke, when the weather cleared up and the sun showed his face between retreating clouds, my relief was unbounded.

Never has that mighty life-giver been greeted with greater satisfaction.

II

Treasure Island
An Encounter with a Shark ~ And a Case of Desertion

After two more days of light and unsettled weather we sighted Cocos, a haystack resting on the northern horizon some thirty miles off. At that moment it stood out very clearly between the neighbouring clouds. The day was fine, and a gentle sou'westerly breeze carried us along toward our destination. By noon we were abreast of the first off-lying rocks, sharp fang-like pinnacles that protruded gleamingly from out of the depth, the whitewashed resting-places of innumerable sea-birds.

As I had been warned that the chart of Cocos Island was not reliable, I determined, after just missing a patch of sunken rocks off the south-eastern point, to keep well away from the coast until nearly abreast of Cone Rock. Then, turning to port, we worked our way against the light breeze into the smooth loneliness of beautiful Chatham Bay.

The water in the bay was clear as crystal, so extraordinarily clear, indeed, that one might almost imagine the surface to be only a plate glass roof over the deeps below, where every detail could be distinguished even down to fifteen fathoms or more.

The bottom of the bay consisted of white sand with large boulders of blue-black coral scattered about. *Teddy,* clearly outlined, seemed to float unsupported above it as if in air. Everywhere the sea was alive with fish of a hundred varieties, most of them gaily coloured as if in carnival attire. Sharks too were numerous, wherever the eye scanned the water.

It had been my intention to anchor in some five fathoms, but the transparency of that sea was so deceptive that when I dropped the hook, I found that we were still in fully ten fathoms of water. If I had taken soundings, we might have gone much further in and thereby saved much labour of rowing, but, having once dropped anchor, we remained where we were, some four or five hundred yards from the head of the bay.

The afternoon being well advanced, we decided to stay on board and spent the remaining hours of daylight in fishing and in watching the strange doings of those queer and fascinating dwellers of the deep. It was like looking into a huge and wonderful aquarium.

The next morning, collecting as many buckets and empty kerosene tins as our little dinghy would hold, I prepared to go ashore and fetch water from the brook at the head of the bay in order to replenish our water tanks. Thinking that I might gather a few coconuts at the same time, I also took a hatchet along.

Cheerfully setting off for my first landing on Treasure Island, I had scarcely come twenty yards away from *Teddy*, when I discovered that I had company, a huge shark following in my wake.

Sharks I regard as repulsive beasts, and I know them to be cunning. They never attack unless it is possible to do so in comparative safety, unless they have a decided advantage over their victim. I did not like this shark following me. I did not like his company.

I pulled harder; he kept pace easily, only the faster I rowed the more determined he became until after a few more strokes, he disappeared under the dinghy.

Knowing how easily the little boat would capsize, I felt fairly uncomfortable but nevertheless continued rowing, until I felt the dinghy lifting slightly on one side, as the shark came up alongside. In desperate excitement I grabbed the hatchet and dealt the brute a slash in the back, which caused him to retreat in such a hurry that he nearly upset the boat after all.

I saw him no more, but a streak of blood in the clearness of the water showed me the course he had taken.

After this incident I never went rowing without the hatchet, and I took good care that it lay handy, especially when I was taking Julie and Tony with me in the boat.

It was only the next day, however, when the dinghy became the cause of another exciting experience.

I had gone ashore at daybreak to look for something to eat, and as there were no nuts on the cocos palms along the beach, I decided to try to find some up the hillside. It was nearly high water when I landed, and the beach, which at low water is dry to the extent of some sixty yards, was entirely submerged.

Meantime I pulled the dinghy well up onto the pebbly bank beneath the verdure, and fastening a piece of good rope to the end of the painter, I tied her up to a tree that grew handy. Whereupon I went on my errand.

The hillside rises abruptly around the bay. It is overgrown with a variety of trees, ferns and shrubs, which are interwoven with a wilderness of tough limbed creepers and fallen trees in every stage of decay. I had to cut my way with the hatchet.

In the beginning I made but slow progress. The ascent was steep, the soil was slippery and the shrubbery proved hard to combat, but as I came higher up, conditions became easier.

Gaining the first hilltop, I found the forest reasonably open, less encumbered with creepers and underbrush. I began to enjoy my exploring.

The vegetation was a picturesque combination of Central American and Pacific island flora. There were tree ferns and palms of many kinds, banyan, pandanus and a hundred other trees, orchids and an abundance of large-leafed weeds.

The morning was still fresh and the air was fragrant with strange and pleasant odours. The birds were almost tame. Large butterflies fluttered about. Absorbed in the surroundings I had long since forgotten what I came for. On and on I went, finding over some new vista opening up for my admiration, another freak of nature to marvel at.

From time to time I would mark a tree with my hatchet to enable me to find my way back. Thus I blazed my trail.

In this manner perhaps two hours had elapsed before my conscience overcame my zeal to explore and caused me reluctantly to retrace my steps.

Returning was easy enough. I followed my own track, and after half an hour's brisk walking came to the hillslope, where an uncertain peek through the rustling foliage permitted me a faint glimpse of the bay and of *Teddy* at anchor.

But I imagined I saw something more. I did not stop to make sure, because that something would not let me. On the contrary, it made me rush down the hillside and through the thicket at a breakneck pace.

Presently I stood under the palms on the beach:

The DINGHY WAS GONE!

So it was indeed the dinghy that a little while ago I saw swinging round the point.

One glance showed me how it had happened; it was just after new moon, the tide had risen to an unusual height; the surf reaching the light craft had set her afloat and then waves and tide had combined to saw off the painter against the sharp edge of a large rock. The rope with part of the painter was still as I had tied it.

Imagine my predicament!

Out there, some five hundred yards away lay *Teddy* at her anchor, with Tony and Julie aboard.

She could do nothing. I had to get on board by my own means, and as soon as possible, while there was yet a chance to overtake the dinghy. Swim there? That would be madness. The bay was simply crowded with sharks; they were everywhere in scores, in hundreds, in thousands. Even in the shallow water along the beach, in less than two feet, their ugly black fins were abundantly evident above the surface.

I might succeed in climbing around to a steep cliff on the eastern side of the bay, but even then I would still have two or three hundred yards to swim, and that area abounded with big sharks.

These and a hundred other reflections flashed through my mind in a fraction of the time it takes to relate them. Indeed, my plans were laid probably in less than a minute and immediately I set about putting them into operation. Fortunately I had a least a hatchet.

The receding tide had already uncovered a portion of the smooth sand. Dry driftwood lay in heaps just behind the beach. In a few minutes I had carried together a respectable pile. I did not have time to worry about the scorpions, which dropped liberally onto the sand from out of the burdens I carried.

Saplings were plentiful and close at hand. I cut down a score and tied them together into a frame by means of rope-yarns; the rope and what was left of the painter admirably served my purpose. Three saplings on either side and four at the bottom made a crate which I subsequently filled with light driftwood, endeavouring in spite of my haste, to place them so as to strengthen the frame. By the time my primitive raft was ready, I was soaked in perspiration. It ran into my eyes and blinded me.

And yet the worst of my labours were yet to come: I must launch this porcupine-like piece of ship-building. It resisted my efforts as if rammed into the ground. I jerked and pulled. I wept and cursed. Several times I was forced to lighten the thing and shove it into deeper water before refilling it and relashing the saplings which served as top beams. And then a sea would come along and shove the whole outfit on to the sand again.

However, success at last rewarded me; my porcupine was afloat and armed with my pole I managed to scramble aboard.

Having a fair idea of the set of the current, I judged that by poling myself along to the eastward, I should eventually get into a position whence the tide would carry me right down onto the *Teddy*.

At the outset things went fairly satisfactorily, although that awful craft made every attempt to fall to bits, but suddenly, without warning, the sea-bottom dropped away. I could no longer reach it with my pole, and I had not come nearly as far east as I should have come.

I tried to row, but that was evidently more than my raft would stand. Besides, rowing did not benefit me in the slightest.

The current carried me slowly on. Sharks were gathering around in large numbers. I could see them even between the branches on which I stood. Twigs kept continually floating away from the raft. They surrounded me in an ever-widening circle.

Teddy drew nearer: it was evident that I would drift by!

Julie was below, possibly asleep.

I began to hail the boat, shouting at the top of my voice, no one appeared.

This was a nice situation! I had certainly come from the frying pan into the fire. If I had drifted to sea in this frigate, my swansong would be sung.

Again I called, howling so loudly that it seemed as though the boughs beneath me rose up in protest and threatened to leave the assembly. Praise and Glory! I heard the dog barking below—how wonderfully sweet that gruff voice sounded!—and then Julie appeared.

She clasped her hands in consternation when she saw me. I was some forty yards from the boat, and I was about to drift by.

However, by great luck, a heavy fishing line lay uncoiled on deck. Tied to one end it had a square piece of wood, on which I used to wind up the line before putting it away. With this the connection was eventually established. And none too soon.

The rest was easy. Pulling myself gently up to the *Teddy*, I caught hold of the main boom end and swung myself on board, just as my frail raft on the slight collision with the boat dissolved itself into a mass of floating driftwood.

My wife had at that time no idea of the reason for this 'queer outcrop of my love of sensation' as she termed it, but I gave her the essentials of the case in a few words. Ten minutes later I had put a buoy on to the chain, hoisted the sails, and we went to sea in pursuit of a runaway dinghy from Ancas Boatbuilding Yard at Arundal.

Following the direction of wind and current as best we could for two hours, we were just at the point of giving up the chase and returning, when at last we caught sight of it. By this time it was half full of water, but fortunately none of the gear in her had as yet been lost.

Before night we were again at anchor in Chatham Bay. The dinghy floated astern, tied up with a new painter. The incident had after all cost us nothing but a bit of rope, a scare and a good deal of work, which latter, I am told, is but wholesome.

However, Spare Provisions was that night overfed in spite of our shortage of supplies.

12

Happy Days on Treasure Island
Into the Pacific ~ Short Provisions and a Long Run

For ten days *Teddy* remained at anchor in Chatham Bay, while I played the happy part of an island king. It was a wonderful time.

I looked at everything with the eye of a Robinson Crusoe; all I saw belonged to me, the waving palms along the beaches, the forests on the mountains, the glittering cascades, the mountain streams and the waterfalls, the beaches and the bays, everything was there to be turned to use for our own convenience. I am afraid I did not spend much time in looking for the treasure, and yet merely to imagine its existence added to my pleasure.

I went exploring to my heart's desire. Some of the little valleys I discovered on my excursions into the interior were of striking beauty. I would stand and gaze entranced at those wonderful revelations of fairyland, those glimpses of a sylvan paradise, and at last I would force myself to bid farewell with a feeling of deep regret.

Tracks showed me that there were pigs on the island, and I had been told that wild cats were numerous, but neither crossed my path.

I have frequently been asked why I did not attempt a serious and systemic treasure hunt. My reply is that in my opinion the visitor who roams idly and pleasantly about stands as much chance as the most laborious and systematic treasure hunter. Furthermore I viewed with distrust the possibilities of success, seeing that some fifty milliard tons of surface are at the disposal of the enterprising gold-digger. Our provisions might not last till I had finished the job.

Unfortunately I did not possess old Cut-Throat's chart—you know the torn piece of yellow parchment with magnetic bearings drawn in blood—neither did I stumble upon any signpost in the way of skulls and crossbones.

But I loved to roam about the shores in the dinghy and look into one or the other of a hundred tidal caves and imagine that just in this one, there in the sapphire and cobalt shadows two fathoms below the sea, yes, just in that cavern behind the rock, there was the cache.

Oh, those lovely caves! How full they were of beauty and marvels and murmurs and mystery!

After all, I believed that the treasure of my dreams was of truer value than any treasure trove. In finding it I would have lost something greater, perhaps even my independence, and unless I had used it—as remarked earlier—for the utilitarian purpose of replacing my pig iron ballast, it might have proved in the end only a curse to me.

Sometimes, in my wanderings I would stumble across bonny little spots amongst the palms where the enchantment of exquisite loveliness and the soft languorous rustle of the leaves overhead seemed to weave a spell so that my one desire was to lie down and just dream away contented hours in the sunshine.

It was a wonderful time.

However, to descend to earth, very little assistance was rendered by the island in replenishing our larder with stores of coconuts, fruit or vegetables. On one occasion I had the temerity to cut down a young cocos palm. We ate the core in the upper part of the trunk. It made the most delicious vegetable dish I had ever tasted. I should have taken some more; the palms were growing much too close in places. I did not do so. Perhaps I had a vague misgiving that after all it was not the right thing to do.

Meantime, fish we had in superabundance. At first I sought the aid of my line, but more often than not a shark would have my catch before I got it on board. After a day or two of this I used the fishing line only to lure the sharks to the surface and then deal them a mortal slash with the hatchet. I had declared holy war on them. By the time we left Cocos, the bottom beneath our boat was strewn with the dead bodies of sharks I had killed in this way. The method seemed to have the effect of keeping others at a distance.

As to catching fish for our table—I resorted to a four-pronged spear. I used to crumble up a handful of biscuits and spread the crumbs on the water; this would bring the fish to the surface in hundreds, sometimes six or eight different species at once. I would then call Julie and ask her what particular kind of fish she preferred for our breakfast, and whether her choice lighted on the pink, the grey, the blue or the black, that particular species always adorned our table. During our stay at Cocos, I became an expert in the art of spearing fish.

In spite of this piscine prodigality, however, our commissariat, small at the outset, kept steadily decreasing, so that at last we had perforce to bid a reluctant farewell to Cocos, this lovely little island of fabulous treasures.

Before leaving, I went ashore with a chipping hammer and cut *Teddy*'s name and station into a rock on the beach, by way of a visiting card—*pour prendre congé**.

When I hoisted my sails in Chatham Bay it was my intention to go to Post Office Bay in the Galapagos Islands. We hoped that if the wind came fair, we might be there in time to celebrate the national holiday of Norway, May 17th, in the company of some of our countrymen who have settled in these islands.

However, the wind kept dead ahead, and since even the current ran against us I soon realised that it would take the best of ten days to reach there. Such a waste of time was unthinkable.

Our provisions were by this time running alarmingly low and I had little faith in the possibility of replenishing our stores in these poorly endowed islands, particularly as I had no money and knew there was no possibility of obtaining any there. If, however, we economised and did not make too prolonged a passage, our provisions might last us until we came to the Marquesas. In view of this, we decided, after a few days at sea, to cut out the call at the Galapagos Islands and sail straight for

* For permission to leave—Ed

Nuku Hiva.

A fresh southerly breeze favoured our decision. As we altered our course to west-south-west, it permitted us to ease off our sheets and make good progress towards our distant destination, some 3500 miles away.

The wind was six points off her head, *Teddy* looked after herself. It was unnecessary to watch the helm. The crew was free to indulge in other occupations. In sleeping, reading, playing cards, painting, mending sails, fishing, cooking and similar diversions we passed the time, whilst—day and night—*Teddy* unswervingly kept pushing ahead. The little spare compass I had placed in the cabin pointed unchangingly west-south-west.

In such fashion three or four days passed. I felt like a passenger on my own boat. Our cooking flourished and turned out countless flapjacks.

However, as *Teddy* worked her way west the wind kept gradually pulling round to the eastward. Our average course became evermore southerly and my journeys to the tiller ever more frequent.

Thus I kept on experimenting with sails and rudder with the object of finding a way whereby the boat could steer herself even with the wind free. Finally I discovered the right principle and thereafter, having properly balanced the gear, I could leave *Teddy* to manage herself on any course, in any wind, with the sole exception, of course, of head winds.

After the first week at sea and until we arrived at the Marquesas Islands, the whole time being taken into account, we hardly spent one whole day at the tiller and this in spite of the changeable conditions we experienced as regards seaway, currents and force and direction of the winds.

Naturally it could hardly be expected that the boat should be always exactly on her course. This, however, caused me no worry. As long as I knew the direction in which we were travelling, I felt perfectly content. If for a while we had come too far to one side of the course, it could easily be compensated when readjusting the balance on the next occasion.

Certainly to make the boat steer herself could only be achieved at the expense of her speed, but this again was more than equalised inasmuch as she kept on sailing incessantly day and night, whether we were awake or asleep. The possibility of meeting other ships in these parts of the Pacific could safely be left entirely out of consideration.

When I happened to wake up at night, I would throw a glance at the compass to satisfy myself that *Teddy* was on her course. A fairly lazy period, this.

Nor was the heat oppressive. Although never far from the equator on this journey, we found the atmosphere at night so cool that it was necessary to resort to a blanket.

On May 25th we crossed the line. Father Neptune arrived in state to administer the traditional baptism. It was intended to be an impressive ceremony, but Tony knocked a piece out of the shaving dish and, thereby, out of the solemnity of the affair. He added further indignity by eating of the wonder-working, if somewhat sticky, shaving substance by pulling the beard off Father Neptune and by demanding immediate surrender of the big razor.

He was awarded his certificate as a pirate.

The mysterious ingredient in the dish, however, certainly did work wonders sooner than we expected.

Tony became more amusing day by day. He evinced a passion for dancing, stimulated by any kind of continuous sound, whether issuing from the gramophone, the coffee grinder, or even from a block in need of grease. Sometimes also he would vocalise lustily to the accompaniment of the records, singing louder and more shrilly than even our tinned Caruso.

As our stock of potatoes on leaving Cocos was found to be sufficient for five or six meals only, it had been decided that this vegetable should figure on the menu only once weekly, so that their consumption might be spread over the longest possible period. It seemed to me that in this manner we would derive the utmost possible benefit from them.

Unfortunately we very soon discovered that the cockroaches were feasting liberally upon our potatoes, and we therefore considered that it was much more advisable that we should dispose of them immediately.

Thus it came about that for more than one month we entirely lacked potatoes, relying for our supply of C vitamins on half an onion a day amongst the whole crew.

These same huge cockroaches had come flying on board in schools during our stay in Panama, that land of large insects, where an ordinary house spider is the size of an average crab. We also housed the smaller brands of cockroaches; such had come on board with the provisions. I verily believe that all the known species of roaches in the world were represented on the *Teddy*, and certainly all seemed to be doing remarkably well.

The big fellows were very fast and they served at least the purpose of affording me some excellent exercise, as I furiously pursued them round the cabin with an old rubber sole.

Days and weeks passed. Our food grew ever more restricted. Though fish abounded nearly always—we caught barracoudas, bonitos and dorados, gaily coloured and sporty, with a spinner—their flesh, so welcome at first, soon palled on the appetite and was eaten only after a considerable effort of will. The tropical fish is much less tasty than the fish abounding in our own native waters. It was much relished, however, by the dog, who waxed fat, while we continued to grow thin.

For some time we were accompanied by three exceptionally fine dorados. By day they would roam about, hunting flying fish. I could not help admiring their lighting-like speed. Even in a fresh breeze, when the flying-fish, leaping from wave top to wave top, would make a hundred yards or more in each flight, the dorados would keep pace with them under water, and always they obtained their prey in the end.

By night they would swim under *Teddy*'s counter, apparently without a motion. The moon was in her second quarter, everything was flooded with light. Bending over the rail I could plainly see the graceful forms of our three followers a fathom below the surface. The smallest of them was easily six feet long.

I had become accustomed to regarding them as part of our larder, and such they would have proved but for my foolish obstinacy.

The day had come to kill one of our fishes. To spear one at night would have been a simple affair, but the probabilities were that my spear would been ruined. Besides, while I desired my bit of sport, my wife wanted her fish fresh from the watery deep. I therefore prepared my heaviest fishing line with a strong piece of bronze wire and a six-inch spinner. The end of the line I tied to the bight of a hand line stretched between the shrouds and the horse. When the fish ran out the line, this contrivance acted as a spring.

Three times I had the biggest fish on the hook, three times I had him alongside, but each time I indulged in the conceit that I should be able to throw him out of the water and up on the deck without assistance.

Of course I failed, because such a feat is impossible. Had I but used horse sense and a gaff I might that day have become the proud possessor of a world record. For that fish was surely seven feet long and decidedly the biggest dorado I have ever seen.

When trying for these fast ocean fish I have always found it a good policy to make it difficult for the fish to grab the hook. I tear it away from them, whenever they make a run for it. In this manner they finally get so worked up that they forget their usual caution and go for it headlong. In the end, I feel convinced, they are quite capable of grabbing the spinner anywhere and at any time, even in mid air. All that is necessary is to get them sufficiently excited.

I think it pays sometimes to study the psychology of the fish. They are, certainly, not so devoid of intellect as is generally assumed. On frequent occasions, when throwing out my line, I have observed that the fish I was desirous to catch would deliberately turn on their sides and direct a careful scrutiny towards the other end of the line. On discovering my presence, they thereupon refrained from even glancing at my fishing gear, no matter how bright and lovely it might shine.

At this stage, the trade wind proved somewhat fickle, ranging in force between a dead calm and a full storm. As a rule our progress varied between eighty and 100 miles per day.

After a careful study of the American pilot charts and other sources of information, I had come to the conclusion that on the best part of this passage I should allow for an average of fifteen miles of westerly current per diem. Adding this to our mileage I calculated that at noon on June 18th we were in 138 degrees, 33 minutes west longitude, and in accordance with this reckoning we now began looking for land.

However, day after day went by and nothing hove in sight. It was only after sailing six days more and after deducting altogether a distance I had allowed for current, that at last the steep outlines of Fatu-Hiva became visible through the sun haze. We were then within five miles of our calculated position.

Presently the Marquesas showed us, one by one, Hiva-Oa, Ua-uka, Ua-pu and Nuku Hiva. The following night found us just off Tai-o-hae. The wind, however, being very light and the night dark, we hove to on the port tack to wait for daylight.

At dawn we put about again and, heading for our destination with a gradually freshening breeze on our beam, we soon made out the rocks—*les Sentinels*—on either side of the entrance to Tai-o-hae Bay.

Just outside I killed a huge shark, spearing him through the head with an old bayonet tied to the end of a boat hook.

The monster had for some time been following in our wake, and I naturally entertained scruples against making my first appearance in a place where I was unknown in such notoriously bad company.

13

The Marquesas Islands ~ Impressions and Experiences

Its black rocky masses rising steeply from out the sea to a height of some 2000 feet, Nuku Hiva, seen from seaward, has a sinister aspect. The island appears more or less flat at the top, the plateau at its highest reaches an altitude of 4000 feet. It is densely wooded. Unlike other islands in the Pacific the coasts of the Marquesas, with one exception, are not fringed by coral reefs.

Three indentations along the southern coast of Nuku Hiva form very pretty harbours, Comptroller Bay, near the south-east point, Taioha'e in the middle and Taioa a few miles further west. Of the three, Taioa is undoubtedly the best harbour, affording shelter in all weathers, but Taioha'e is the principal village and port of entry.

The Spanish navigator Alvaro Mendana de Neyra was the first white man to call at this harbour when he discovered this group of islands, AD 1595.

Arriving just as the war drums—shaped like huge hour glasses—summoned the warriors to a feast on slain enemies, Mendana thought the natives a ferocious tribe. When Captain Cook called at Nuku Hiva one hundred and fifty years later he found the Marquesans living in the same state of primitive barbarity, in which they were destined to remain for yet another century, until in 1842 the sovereignty of the Marquesas Islands was ceded to France by a treaty with Admiral du Petit Thouars. Now it is long since the French put an end to civil warfare and cannibalism; whether this is due to their firm rule, the influence of the gospel or the introduction of bully-beef by the traders, I shall not argue.

However, I was told that only twenty-odd years ago the authorities of Hiva Oa came across a case of murder and cannibalism, some exultant heirs feasting upon the remains of an unpleasant relative.

As we approached the anchorage I became aware of two gentlemen paddling towards us in a pirogue. They were, as I subsequently learned, the assembled medical authorities of the Marquesas group, coming out to give *Teddy pratique*.

When we had anchored and lowered our sail, they came on board. Dressed in pyjamas with short pants, these twain appeared considerably less ferocious than one would expect Marquesan medical men to look. They proved to be Dr Rollin from Hiva Oa, well known as a writer and scientist, and Dr Quere, the local medical man.

When the formality of examining my papers began, it transpired that neither of them could furnish the date. Indeed, this was charming! Here, at last, we had come to a place where time was of so little importance that not even the authorities knew the day's name. I felt amongst friends, and I suddenly took an immense liking to Nuku Hiva and its inhabitants, and a liking which was certainly in no wise diminished by my subsequent experience here.

For lunch that day we resorted to our last tin of meat. It was the only meal that was left on the ship. Truly a sufficiently close shave.

Still we could find no fault with our domestic economy. Forty-five days had passed since we left Cocos Island and seventy-one since our departure from Balboa with provisions calculated to last us for one month only.

But neither my wife nor I shall ever forget the sweetness of that first orange, brought aboard by the doctors, the first other human beings we had seen for more than ten weeks.

The resident Commissioner, M Triffe, came on board in the afternoon, bringing his wife and several young ladies. He presented us with baskets full of fruit and vegetables and heaps of green coconuts to drink. We became great friends. I believe that while at Nuku Hiva we had our meals in his house as often as we had them on board, and when I was away on excursions, Julie and Tony were his guests.

I remember with distinct pleasure some of the delicious meals we had in that hospitable home, and yet, with the sole exception of the French wines, there was not a thing on the table that had not been produced on the island. I have never tasted coffee to compare with the coffee grown in Nuku Hiva and prepared by Mme Triffe.

Taiohaʻe is very lovely and very quiet. Frequently in the daytime I marvelled at the strange stillness of the place. A semi-circle of towering forest-clad mountains looked dreamily down upon a scene of listless languor. The scattered houses of the settlement showed sleepily among motionless foliage. Deserted lay the white coral road shaded by graceful palms and puraos. Deserted also was the impossible little pier below the residency. The soft rattle of the oars in the tholes, as I rowed leisurely ashore, the muffled wash of the surf and the grating of the dinghy's keel on the sand merely seemed to intensify the profound stillness. But presently a bevy of Marquesan children would spring into life from somewhere and break the spell. Birds would start to twitter, dogs to bark and Mme Triffe's rooster to crow, blending with the voices of the children splashing noisily about in the surf. However, nowhere in the Marquesas are children very numerous.

A businessman would find—perhaps—that the village of Taiohaœe is stagnating. He would be quick to notice that the better part of the scattered houses along the single bayside street are in sad want of repairs. He would discover that the club—*le Cercle International*—where formerly a score of white men in spotless ducks gathered together after sundown for their absinthe hour, has ceased to exist. He would be informed that there are not sufficient white men left to uphold that venerable establishment, and—being a quick observer—he would see that the few whites who remain are dressed according to their work and circumstance without any apparent regard to racial distinction. The era of spotless white ducks has passed.

Then, returning to the schooner that had brought him, our businessman would engage in conversation with my friend, the captain, to whom he would open his heart, giving as his opinion that the white residents were a luckless lot, and regretting that he had wasted his time in coming to such a God-forsaken place. 'If ever this place had any ambition,' he would declare, 'now it has none.'

'Oh, well!' I hear the captain's soothing voice reply, 'What can you expect? Trade in these parts has shrunk to an insignificant figure, due partly to the sadly reduced buying power of copra, but more particularly to the rapid decrease in population. We are running the only ship left now in this trade; our competitors have all pulled out, and I don't know that it pays us to carry on. Yet a few years ago this was the best paying trade in the Pacific.'—Not a very gratifying reply to our progressive friend.

He would look at the profuse vegetation and the deserted street, and finding no answer to the puzzle he would turn and go below, where he would more justly vent his vexation on those hateful little day flies of Nuku Hiva, the poisonous nonos.

Still, our friend would be right, Taiohaé has no ambition left and scarcely any hope. It is moribund—doomed, unless perchance it is someday revived by the Chinese.

But in spite of the resignation which seems to brood over the place and the minds of its inhabitants, it has preserved an indefinable lure of romance, an air of sweet sentimentality which strongly appealed to my senses. Who could resist the charm of those early night hours when the land breeze, heavy with fragrance, brought freshness down from the heights, when vague sweet melodies floated upon the air, distant song, originating, perhaps from a group of Kanaks who had lit bonfires on the far side of the bay to attract fish.

Then join me for a walk, my friend, along that length of coral road. Do you see those scattered lights glowing in the green thicket? Do you hear the faint rustle of the palms overhead? Do you notice the brilliance of those stars?

As we go along, comely native women will emerge from out of the darkness by twos and threes, smiling and ogling us and, perhaps, wooing us with a strain of soft laughter—and leaving to the night air, as they pass on, a bewildering perfume of palm oil and frangipani blossom. Wreaths adorn their rounded necks and a large flower, tucked above one ear, indicates the fact that so far they are unattached.

Do you not think, my friend, that the fragrance of the breeze mingling with perfume of those frangipani blossoms, the shadows of the tropical night, the soft laughter of alluring maidens, the harmony of distant song, the murmur of the surf, lend charm, romance, nay enchantment to this nocturnal scene? I at least do!

The rapid depopulation of these fertile islands is a deplorable feature. Two generations ago the Marquesans are stated to have numbered 60,000 people. Scarcely 1200 natives can now be numbered, and within another thirty years, so the resident Commissioner informed me, this fine native race will in all probability have ceased to exist.

The Marquesans are well built and pleasantly featured. They remind me, more than any other Polynesian race, of the Maoris of New Zealand, and they exhibit the same fine traits of character.

During our stay I went pig-hunting in the Happa Valley, familiar to most people from Herman Melville's great book *Typee*. Not so many years ago there stood in the valley a village of some four hundred inhabitants. Within twelve months all but two had died, killed either by tuberculosis or by smallpox. Of the village itself nothing is now left but the mouldering 'pae-paes,' the stone platforms on which the Marquesans used to build their houses. Even the 'pae-paes,' however, are not readily discovered in the dense thicket.

A variety of game abounds in the islands, including cattle, goats, pigs, and fowls, turned savage after the disappearance of the original owners and breeding freely in the bush.

Butchers' bills were unknown in the Nuku Hivan households. Once a week a man was sent into the mountains with a packhorse to shoot for the requirements of the house. Oranges and many other kinds of fruit and vegetables grew wild and in profusion. There was no need for anyone to starve.

Taioha'e boasts a prison and also a prisoner. The latter often came along in a pirogue to show his mastery of the English language. His vocabulary comprised three words, 'Goo-bye,' 'mister,' and 'aw-right,' which he used indiscriminately and variedly, and by means of which he managed to maintain a voluble conversation to the great astonishment of other natives who might happen to come near. The prison was never locked, and the prisoner came and went as he liked, his sole duties consisting in keeping the prison clean and doing a few odd jobs about M Triffe's house. Apparently he was no worse off than anyone else, being in fact better housed than the majority of his people.

Two other ships called at Taioha'e while we were there, one the auxiliary schooner-yacht *Zaca*, belonging to a great banker from San Francisco, and the other the trading schooner *Tereora*.

The American banker gave us a dinner-party and, on leaving, presented us with a good many useful things. Particularly welcome were two large tins of milk powder for the boy; our stock had run low and there would have been no other opportunity for replenishing it.

Although the anchorage at Taioha'e is sheltered against the prevailing easterly winds, which leave only a gentle swell at the head of the bay, a south wind occasionally blows right into the harbour, bringing up a heavy surf on the beaches and making ships tug at their chains.

Then, mariner, beware!

To be sure, the holding ground is excellent. The changing tides, however, together with the diurnal change of the wind in the bay, cause ships to swing in a continuous circle round their anchors, eventually coiling up the whole length of chain around the flukes until the chain stands perpendicularly up and down. Then, of course, with the next rise of the tide, the ship herself will lift the anchor out of the ground and start dragging.

That is exactly what happened to *Teddy* the day before our departure. Fortunately the *Tereora* was at hand. I was just about to hoist my sails and slip the cable, when the captain of the trading schooner discovered my plight and shouted to me to hang on for a while.

Presently a reef boat, manned with five or six athletic Kanaks, came rushing to my assistance with an anchor and a heavy rope. When I tried to express my gratitude to the captain afterwards, he waved it off, declaring that he had done no more than any sailor would do for his fellow. Which, I think, was a remarkably fine way of proving that he was a true sailor man.

Before we left, some of our lady friends presented us with three huge bags of fruit, coconuts and vegetables, a roast and about ten dozen eggs. Little likelihood of starvation on our next voyage!

Fishwives at Lisbon

Teddy at Santa Cruz

The Mate

Making discoveries with a sextant

The Skipper and 'Spare Provisions'

Teddy at Cristobal

Teddy at Balboa

Tai-o-hae

Marquesan visitors

Papeete

The Tambs family and *Teddy* in Papeete

Moorea, Papetoia Bay

Tony and friends at Raiatea

Fishermen of Bora-Bora

The Chief of Bora-Bora and the Chieftainess

Tony fishing

Tony's 'Big Fish'

The Mate

New Zealand earthquake

Main Street in Napier

Start of the Trans-Tasman Race

Looking forward to leeward during the race

Teddy ashore in the Bay of Islands

Teddy on the slipway in Auckland

Leaving for Tonga

Skipper and Mate

Castaways

Wreckage from the *Teddy*

Summer Isles of Eden

We weighed anchor at daybreak on Monday, July 7th. The government launch towed us out of the bay. Once we were outside, we set the sails, and, bending to a stiff easterly breeze, *Teddy* was off. We passed close to leeward of the island of Ua Pou, with its towering peaks and steep shores resembling an ancient castle on the Rhine.

Before nightfall the outlines of Ua Pou, looking at the time the very image of Salzburg in Austria, had faded away in the mist astern. We had left behind the Marquesas Islands, populated today by the scanty remains of a lovable people resolved to die, but once inhabited by a stately tribe of fierce, proud, happy and healthy cannibals.

The trade wind blew fresh and we made good time. On Thursday night, finding ourselves according to my calculations within a distance of some twenty miles from the atoll of Manihi, we hove to, intending to make passage between the first group of atolls in daylight. I did not care to navigate amongst the low islands of the Tuamotu archipelago in the dark without being certain of our exact position.

However, the next day fell calm. We made very little headway, and night found us, probably, about midway between Manihi, Takaroa and Takapoto. All day long I had been looking for a sign of these islands, from time to time climbing the rigging to obtain a wider view, but nothing had hove in sight.

Having sailed four hundred odd miles with nothing but an approximate compass course to give us our longitude, we might easily have been fifty miles east or west of our calculated place. I had, therefore, a variety of courses to choose from, any of which gave my nautical problems a different aspect. Meantime I could do nothing while the calm lasted.

At midnight, however, a breeze came up from the east and, as it freshened, it brought to my nostrils the unmistakable smell of land. I knew it from my Marquesas experience—that unmistakable odour of island wood fires. The entrance of this new element was gratifying. It was clear that if we had islands to the eastward, we could not have any immediately ahead. Accordingly we proceeded on our course.

When day broke we sighted an atoll, at first a single palm that showed above the horizon ahead, disappeared and again showed up, then another one, a group, several groups…

They stood out ever more distinctly. Presently appeared the undergrowth, the reef, the surf, the white beach. It proved to be Arutua, the very place I had been striking for; the lay of the neighbouring islands made its identity evident.

As usual I endeavoured to conceal my surprise under a becoming mask of indifference, implying that I had never seriously doubted the reliability of my navigation. On this occasion, however, my air of nonchalance failed to convince my wife. She laughed whenever she looked at me.

Sailing close along the reef of Arutua we had the benefit of smooth water until the wind showed signs of fading out, when I found it prudent to make for the open sea. I had read so much about the treacherous currents among the islands of the Low Archipelago that I did not want to be caught close to a reef in a calm.

Clear of the island, we met a nasty southerly sea and presently found ourselves becalmed. The boat, now deprived of the steadying pressure of the wind, was being tossed about by turbulent waves in such a manner that only experience and care saved me from being thrown overboard. I could still see the palms of Arutua in the distance. It was towards dark, and I stood on deck wondering if it were not better to take the mainsail down, when suddenly something happened. Flying blocks, sagging canvas, loose lines and wire ropes whizzed about my head and crashed against each other in a most disconcerting fashion.

In an instant the whole rigging had become a confused mess of mutinous parts which individually and collectively strove to smash my cranium. The shrouds alternatingly hung in huge bights or tightened fiercely with a singing sound as the rigging tumbled from side to side with the frantic tossing of the boat. Within a few seconds the mast wedges were chewed to pulp and the collar was torn to shreds.

The heavy mainsheet blocks tore madly from end to end of the horse rail amid a hellish rattle and shower of flying sparks.

A desperate, bewildering moment!

But then I saw: the forestay had carried away!

This was a case for quick action if I wanted to save the rigging. Down came the sails.

The trysail sheet tackle lay handy. With it I temporarily secured the forestay to steady the mast, and then I proceeded to muster my resources. They appeared somewhat inadequate. However, I found a fairly suitable piece of heavy wire rope and sufficient seizing wire to lash it on to the broken end of the stay. The night had settled before I got properly started on what, in the darkness and the frightful turmoil, proved a very trying job. When at last, about two o'clock in the morning, it was finished, I was by no means pleased with the results of my toil. However, my adjustments stood up to the rather severe strain they were put to before we finally reached Tahiti.

Sunday morning found us drifting about south of Rangiroa. Palm-clad islets showed up here and there. The sea was gradually settling down. From time to time a long swell lifted the white teeth of the surf into vision. A school of huge whales playing lazily about in a light blue, sunny sea.

In the afternoon a squally wind sprang up from the north-west, and as it gradually pulled around by west, it developed into a hard blow from the south. Shortly after nightfall the jib blew out at the leeches. At midnight I was compelled to double-reef the mainsail and an hour later also the staysail. But even then it was hard sailing on defective rigging.

We worked laboriously south-west. Monday at noon we passed close to leeward of Makatea, an island of upstanding coral where a French company very successfully exploits rich deposits of phosphate by means of contract labour from Annam*.

As *Teddy* shipped much water on this part of the passage, a good portion of which came rushing down into the galley through the gaping hole around the mast, thus flooding the cabin floor, we did not enjoy much comfort. Furthermore, pumping is not my favourite diversion, although my labours at the pump, occupying a quarter of each hour, might lead one to think so.

However, the weather gradually improved and when on Wednesday morning we sighted the little atoll called Tetiaroa, the wind was light and the sea about to subside. To the south-west the rugged peaks of beautiful Moorea thrust their summits through lofty clouds, and presently the veil lifted to reveal to the eye the manifold charms of laughing, much-beloved Tahiti.

The day was perfect. Lazily we drifted over a sunny sea amidst scenery which the eye could never weary of beholding, pleasing, colourful, lovely, and yet of majestic grandeur.

When a blazing sun set behind the black outlines of Moorea, we were almost within sight of our destination. The sound of the swell breaking on the reef was plainly audible. To attempt an entrance would, however, have been folly. In these latitudes night follows right on the heels of day, and the passage through the reef into Papeete harbour has a bad reputation. We hove to off Venus Point.

At daybreak we bore away for Papeete and before noon dropped anchor in the pretty little port amidst a handful of trading schooners. Stern ropes were brought ashore and made fast to trees on the shoreside promenade, and presently *Teddy* lay snug and safe, stern to the embankment, a gang plank connecting our boat with the smiling little town of Papeete.

---

* Today, central Vietnam—Ed

15

Idle Reflections on Paradise Lost
And some of an Idler's Experiences

Considering that Papeete is the largest town between Honolulu and New Zealand, the metropolis of an immense area of scattered islands, its appearance is not very impressive. Its population numbers about 4000. It is a ramshackle collection of just that type of houses with which the white man enjoys destroying the native beauty of all foreign lands. As usual, the buildings of the business blocks are particularly horrid, though even in the residential parts of the town I found nothing to shake my conviction that no architect has ever visited the island.

Meantime Nature is rich and generous, hiding the monstrosities of local builders under cascades of flaming bougainvillaea, behind ornate palms and ferns, behind hedges of blooming hibiscus and groves of prolific bananas, and lavishly shading the ensemble under giant flowering puraos and acacias, or the graceful plumage of the slender, ever-present cocos palm.

Thus in spite of the total absence of architectural taste displayed by present-day Tahiti, the general effect is often exceedingly pleasing. It is deplorable that even in the countryside the wholesome native palm leaf and bamboo dwellings, which harmonize so perfectly with the surrounding landscape, are gradually being replaced by shacks of board and corrugated iron, offensive to the eye and injurious to the health of their occupants.

In the days of the early explorers, Tahiti was famed as a very Garden of Eden. If today it is no longer paradisiacal in all aspects at least it is, and will ever remain, a fairyland of indescribable beauty. Like a gem of brilliant colours set in the coral-fringed emerald of the lagoon, it rises out of the deep blue waters of the Pacific, a fertile tropical dreamland of rare scenic grandeur under an ever serene sky. The variety of scenery is marvellous, ever changing, ever revealing charms of different nature, from the leafy luxuriance of the palm-bordered beaches to the stern magnificence of lofty mountain crests. Far removed from the busy centres of the world, fanned by the trade wind, gentle, generous, beautiful, Tahiti is indeed an island where the weary may find rest and the melancholy or embittered may find contentment.

As in so many other parts where Chinese are admitted without restriction, the industrious yellow people are slowly ousting the rest of the population from all productive activities. In Tahiti all trades are carried on by Chinamen and what yet remains to the whites of commerce seems to be gradually passing into their hands. The vanilla trade, which is of considerable importance to the island, has long been a Chinese monopoly. Now they have cast their eyes on the copra trade and connected lines. Already the majority of trading schooners are owned or chartered by Chinese merchants, who also run the majority of stores and restaurants on the island. To the casual visitor this state of affairs is not disagreeable. The Chinese supply the market with commodities at reasonable prices. The merchants are polite and honest and the tradesmen work cheaply and well.

Yet it seems to me that Tahiti offers a striking illustration to those who predict that the era of white supremacy is rapidly drawing to an end. Already, on festive occasions, one sees as many Chinese flags in Papeete as one sees tricolours.

The native Tahitians do not worry about this development—they are only too ready to leave tedious toil to anybody who cares to endure it, while the majority of white people seem to drop all pretentions to such disturbing qualities as business energy and initiative immediately they set foot on Tahitian soil. Nor do I blame them. A similar inertia possessed me also.

I had come to Tahiti with the honest intention of flooding the world with literary efforts. I would write novels in a South Sea setting, I would write essays on burning social questions, I would, oh, well, I would at least earn sufficient money to do away with financial worries for a long, long time.

As it happened, those wonderful literary creations failed to come into existence. Life in Tahiti is too easy, too pleasant. Tahiti is an island for play and for *dolce far niente**. I did not touch a pen unless I had to, that is, when our funds had disappeared altogether. I would then compose a newspaper article and hasten with it to the Post Office, even if it were weeks before the next mail left. Having registered my manuscript, I would next send a telegram to the paper, stating that I had mailed an article and asking for the honorarium to be cabled to me immediately. The money for the postage and the telegram I usually borrowed from the Norwegian Consul. In justice to the paper I must add that they never 'let me down.'

Thus I enjoyed plenty of leisure for roaming about the island, for associating with interesting people we met, for swimming in the lagoon, or fishing with the natives on the reef at night by the fantastic light of numerous torches made of dried palm leaves.

Part of my time, of course, was occupied in keeping the boat in trim and repairing the rigging; sometimes circumstance forced me even to do some writing, while my wife looked after the interior department of our little kingdom, kept the cabin spick and span, and attended to the wants of our son and heir, 'Te Arii no te Moana'—a prince of the sea—as a Tahitian girl smilingly called him in passing.

Also, while in Tahiti, I had *Teddy* on the slipway to overhaul her bottom. Imagine my surprise, when, as she was raised out of the water, I saw that more than half of the false keel had disappeared. The missing part was a piece of heavy timber about ten feet by two feet, reaching from the stem to the iron keel amidships. I had never suspected the loss, and I have no idea when or where it had come adrift. This involved further expense.

However, the local shipbuilder, Steve Higgins, repaired the damage

---

* Pleasant relaxation in carefree idleness—Ed

very neatly and at a most obligingly reduced price.

Meantime I believe that the absence of the forward part of the keel must have had something to do with the boat's ability to steer herself, because, in spite of the experience I had gained in this respect she has never since been so willing to keep a steady course with tiller lashed as on the passage from Cocos Island to Nuku Hiva and Tahiti.

Our days passed pleasantly. Rising early, we often went to the market, which was open only from five to seven. Here all Papeete came to purchase its requirements of victuals. Meat and fish could be bought nowhere else. Thus the market had developed into a meeting place, and in the freshness of those early morning hours it was delightfully thronged with happy people in light and gaily coloured clothing. The buyers were mostly Tahitians of the fair sex and the vendors mostly Chinamen. The stands abounded with produce of great variety; fish of a hundred species, shapes, sizes, and colours, meat, poultry, vegetables and the profusion and variety of fruit which only the lavish soil of a tropical island such as Tahiti can produce.

Every morning at eight a young Frenchman came to fetch Tony. In return for a moderate fee he and his native 'wahine' minded the boy until noon, when he would bring our son back to the boat. The Frenchman disposed of his duties in a very sensible manner. Bringing the boy to his place, he would undress him—to save soiling his clothes—and let him play around with a score of other naked children of brown or yellow colour in the palm groves or on the sunny beach. Amongst these youngsters Tony, although barely able to walk, was treated with great friendliness and consideration. Whether our Frenchman had instructed them or not, I do not know, but to them he surely was 'Te Arii no te Moana.' He could not have been better looked after by his own mother than by those Chinese and Tahitian children.

Sometimes, when we were to attend a dinner or some similar function, we would leave Tony with our Frenchman even at night, feeling perfectly sure that the whole neighbourhood would take care of him.

Kipling has said that 'East is East and West is West and never the twain shall meet.'

With the agency of Tony that meeting was effected.

One night in particular I had a rather pretty experience of this kind.

Coming to fetch Tony from his nurse and finding nobody at home, I strolled into a Chinese store intending to make enquiries. I stopped on the threshold. The scene that revealed itself to my eyes in the light of a single kerosene lamp hanging from the roof was too pleasing to disturb. On chairs around in the shadows sat six or eight Chinese mothers, their otherwise plain yellow faces indescribably beautified by the radiant smile of motherhood. In the middle of the room sat another Chinese woman smiling happily upon a group of children in front of her. In the centre of this group, where the lamp cast its brightest light, stood a very white-skinned and very eager little boy talking to his Mongolian playmates in a language I did not understand. They seemed to comprehend and even answered or, in their turn put questions to him, took his hands, danced and laughed.

Tony's face was turned towards me. Suddenly he stopped in his play, and as it dawned on him that it was I who stood there in the shadow against the doorpost, his pretty little face underwent a sublime change from the mere happiness of enjoyment through finer phases of emotion to the divine ecstasy of recognition. At last with shining eye and a little sob of happiness he stumbled towards me with outstretched hands, and as I picked him up he squeezed me again and again with all the power of his plump little arms. I have never felt a moment of greater happiness.

I looked around me. Those Chinese women smiled at me with kindly faces. East met West in the kinship of parentage.

It has been said that the intense struggle for life has rendered the Chinese incapable of emotion. Why, then, was it that some of those mothers furtively wiped their eyes?

16

Dogs and People at Tahiti

Tahiti is a volcanic island consisting of two separate mountainous bodies almost circular in form and connected by a narrow neck of low land. The main island—Porionuu or Great Tahiti—is some thirty miles in diameter, with mountains rising to more than 7000 feet in height, while Taiarabu—Little Tahiti—is just about half as wide across while its summits extend half as high as those of the main island. A narrow strip of flat and fertile land runs all round the island, and a good road encircles the main part of it.

The coast is sheltered by a coral reef at some distance from the shore, forming a navigable channel inside. At Papeete this channel widens out to make a fairly spacious harbour.

As the high mountains shelter Papeete from the easterly trade wind, usually only light airs and calms prevail in the port. Occasionally, however, strong northerly gales drive a heavy surf over the reef and render conditions none too safe for vessels in the harbour. As the weather remained fine almost continually during our stay at Tahiti, we had rarely reason to worry about the safety of our boat.

Teddy was generally left in the care of Spare Provisions, hatches open and the gang plank invitingly laid to step aboard. In spite of these obvious facilities, we never entertained the slightest fear that anyone might take advantage of the situation, although the dog was seldom on board and indeed rarely to be seen. Like ourselves, she would roam about and explore, but in her particular case I fear it was the garbage bins of the neighbourhood that attracted her special curiosity. Whenever anyone came near the boat, however, she would suddenly appear, eyeing attentively the cause of her suspicion and growling warningly if that cause should happen to step on or even touch the gang plank in passing. Otherwise Spare Provisions was a very peaceful animal. She never picked a quarrel with another dog and she always gave cats a wide berth. There was a big brute of a dog that tyrannised all other dogs in the street and which seemed to have taken a particular dislike to our canine guardian. Spare Provisions being the faster runner of the two did not fear her big enemy at all, in fact she completely ignored its presence, unless the other, infuriated by such contempt, would rush at her, when Spare Provisions would hurry back to the boat. But once she reached the gang plank, she would sweep around and face her foe with a snarl—the very picture of an angry wolf—while the other, taken aback and lacking the moral fortitude of the defender, would come to a sudden stop, look uncomfortable, and then slink away, continually glancing over its shoulder to see that it was not being pursued.

I do not believe that any foe, man or beast, could intimidate Spare Provisions in the defence of what she considered to be her home.

She fared well in Tahiti. There was always plenty of food for her and a succulent bone to round off her meal. The only fly in her ointment was Tony. He was as fond of bones as the dog was herself, and if we did not prevent him, he would crawl under the table and, by fair means or foul, seize the dog's bone in spite of her snarling protests. Sometimes a regular tug-of-war ensued. Pulling at the bone with one hand and with the other pushing the dog's nose away, Tony would tug and jerk and push and pull till he finally succeeded, when my wife would step in. It then became a matter of soothing a very offended little boy, who had been robbed of the prize he had won in fair battle.

However, occasionally it happened, when I was on deck, that the dog would summon me with loud complaints about unfair treatment, and, coming below, I would find Tony lustily gnawing at the dog's bone, whereupon I would be confronted with the difficult task of restoring the dog to her rights, an operation which usually could not be carried out without conflict.

The Tahitians are very fond of children. Once or twice I was asked in all sincerity by Tahitian women if I would not give them my boy. Such a request is not at all astonishing in an island where it is customary amongst the natives to give away some of their own children and to adopt others. I am sure that, had I asked, they would have given me in return one of their own children, even their best beloved. Meantime I did not entertain the suggestion. In other parts we might later visit, where this custom of exchanging children is not in practice, it might give rise to unpleasant comments if the colour of my child differed too widely from my own.

And yet, how lovable are those natives! What a laughing, compassionate, light-hearted, unreflecting, irresponsible and thoroughly delightful people are they!

In their happy-go-lucky way they let each day carry its own burden of joys and sorrows. Tomorrow new flowers would bloom to be bound into a wreath for a man, for a girl, flowers to be worn around the neck or in the hair until they faded and were thrown away. How truly significant of their own minds were those flowers!

Often, when a trading schooner was about to leave, have I seen girls on the wharf crying their hearts out in grief at the departure of their lover, sobbing and waving until the ship had disappeared in the distance. Then they would wipe their eyes and turn around to greet with a smile a new friend, and presently the two would walk away hand in hand, happy as two children. The departed lover was forgotten. The natives of Tahiti undoubtedly have many virtues, but stability is not among them.

There is no rule without an exception, and such an exception we found in the person of Terieroo a Terierooiterai, a Tahitian gentleman of standing, chief of Papenoo and highly esteemed counsellor to the governor. He was a man of sterling character and outstanding ability, but at the same time gentle and kindly, a pleasant companion and an engaging host.

He invited us to dinner at Papenoo, a dinner prepared in true Tahitian style. The diners wore crowns of fragrant flowers and sweet-smelling wreaths also adorned their necks. Green coconuts served the cooling drink. It was a most magnificent meal.

Some of the items of the menu I may have forgotten, but I remember such delicacies as raw fish with miti sauce, lobster, cold fried fish, a roast pig, roast chicken, roast kid, bread-fruit, yams, sweet potatoes, cooked bananas, dessert and no end of fruit.

All cooking was done in the traditional native way on hot stones buried in the ground. Although there are many ways of preparing food which are more modern, there is certainly none that is better.

I have never seen more attractive faces than amongst the native girls of Tahiti. And they do not overdress, as do the women in so many of the other islands. Outside the town of Papeete, where certain stupid rules of 'decency' must be observed, a 'pareu' generally satisfied all female ambitions as regards dress. I do not know that I have seen women more becomingly clothed anywhere. The 'pareu' is merely a piece of brightly coloured calico, some two yards long, which is bound tightly around the breast and reaches to the knees. It is the common dress of the men also. The men, too, are of fine physique and pleasant features, happy in nature like the maidens, always ready to laugh and to play.

In the Tahiti of today, the half-casts are gradually outnumbering the pure natives. Unfortunately for the offspring, the Tahitians do not seem to mix too well with foreign races, the half-cast, although often of striking physical beauty, generally being weaker in body than his parents and lacking resistance against diseases. Many of them suffer from tuberculosis, as do the half-cast Maoris of New Zealand.

Strangely enough this does not apply to half-bred Chinese. While rarely overburdened with personal charm, the children of Tahitians married to Chinese are generally strong and healthy.

In Papeete one meets a variety of strange and interesting people. I do not refer to the average tourist or the established residents in general, who are, like the majority of 'decent' people, most uninteresting, but to others, who were not so easy to classify. They were men from all corners of the globe, form all strata of society, from the most varied professions, who had drifted into this Rome of the Eastern Pacific at some stage of their adventurous careers, men who possessed individuality and broad mindedness resulting from knowledge acquired by experience.

Also, since the days of Gauguin and Pierre Loti, Tahiti has become a meeting place of men of arts and letters.

It is one of the privileges we enjoy on this cruise, that we are free to choose our company without regard to their social station. Thus I have rubbed shoulders with governors and boot blacks, admirals and stokers, multi-millionaires and beach-combers, men of the highest culture and savages. Thus I have mixed with people of many creeds and colours and found pleasure in their society by virtue of their common humanity. For this human element I search as diligently as did old Diogenes for an honest man.

In Tahiti one does not need a lantern to find mankind.

17

How I Faced August Authority and subsequently Proved Myself a Pirate

Having come to Tahiti with the intention of staying for some time I thought it would be wise to observe conventional rules of courtesy. Therefore I mentioned to the Norwegian Consul that I should like to pay my respects to the Governor, a matter which the consul promised to arrange.

Now, if anybody thinks that it is an everyday function to call upon the Governor of Tahiti, he is mistaken. Personally I did not expect any fuss in this connection and it was only afterwards that I learnt how elaborate were the proceedings. It proved that the consul had to file an application on my behalf with the *Chef du Cabinet du Gouverneur*, from whom he had in due course received a reply appointing a date a fortnight later.

The Governor proved to be an excellent gentleman and interesting to talk to. He was, I learnt, one of the 'strong men' of the French Colonial Department, a *gouverneur de premiere classe*, come to reorganize affairs in these grossly over-administrated islands. It struck me that there were a few things which could be simplified in his immediate surroundings.

The French authorities are very tolerant. In Tahiti everyone may lead a life to suit himself. The officials, although numerous, are as peaceful as snow on a mountain slope. Only you must not stir them to activity. If you do, they will descend upon you in an avalanche of paper and red tape, and then it is impossible to predict the outcome.

And, although there are so many officials, each is a very important person. Whatever happens, you must never forget this vital fact. It was in this respect, I think, that Rivnac failed, and consequently he was harassed by a number of offended officials. The Bohemian proprietor of the almost historic Hotel Tiare, Rivnac had on one occasion obtained permission to hold a dance at his establishment. He was fined next day for failing to procure a permit to have music! Rivnac's adversities at least did not lack a touch of humour.

I, too, happened inadvertently to offend an official and was consequently kept busy for a week writing newspaper articles to pay the expenses of my default.

The trouble started, I have reason to believe, when I omitted to pay a call on the *Capitaine du Port* on our arrival. Surely I had no intention of hurting anyone's feelings, but I was innocently ignorant of the existence of so important a person in so unimportant a port. I thought that I had satisfactorily discharged all duties of courtesy towards the harbour department, when I had shaken hands with the pilot, who, I judged, might easily have been in charge of it all.

When, at last, I became aware of my neglect, I did my best to reconcile the offended official; I did this—I am sorry to confess—rather with an eye to the possibilities of lenient treatment in the matter of harbour dues, than from any sense of duty. However, I bungled things by telling the truth, and when I took my leave it was evident, from the frigid politeness of the *Capitaine du Port*, that the harm was done, and done irrevocably. Presently the bills began to come.

Seeing that I was in for it in any case, I thought that I might as well extract some amusement from the situation, and, being of a naturally reckless disposition, I would now and again invade the lion's den and pull his leg in polite fashion. He rarely saw my points, to be sure, but he was suspicious of my motive, and as he took himself and his restful little port very seriously, I do not think that my friendly calls made him love me any better.

On the day before our departure I went to the harbour office to get my clearance. Although I knew that I had paid in full all the usual charges, I asked the port captain to make sure that nothing had been overlooked. That caused his cup of annoyance to overflow. Exasperated, he shouted: 'You have paid no more than any foreign vessel is supposed to pay,' which in a sense was true. Because we had been spoilt in other ports by enjoying reductions or complete exemption from harbour charges, I had no right to expect a similar privilege in Tahiti, even though it is surely not customary in France or any other civilised country to charge harbour dues to foreign yachtsmen.

'Very well, then,' I retorted, a little ashamed of myself, 'As there is nothing more to pay I want my clearance. I intend to sail to-morrow.'

'You shall have it,' he said grimly. 'I shall send it on board in the morning.'

Thereupon offering him my hand—which he pretended not to see—and wishing him all sorts of luck and happiness, I departed, truly sorry that I was leaving behind a very hurt *Capitaine du Port*.

The afternoon and the remainder of my Tahitian money we spent in buying additional stores and in paying farewell calls.

It was my intention to leave about nine o'clock in the morning. In order not to make the departure too lengthy a performance, I let go our moorings, picked up the anchors and passed a single rope on board our neighbour, the French warship. Thus, to get underway, it was only required to cast off the rope and hoist my sails.

In the meantime friends had been gathering under the grand old trees which cast their shadows upon the Boulevard Bougainville or whatever may be the name of the shoreside street in Papeete. My ignorance here is shameful, considering that we had had our moorings made fast to those trees for more than three months.

When we came ashore we found ourselves surrounded by a truly cosmopolitan crowd, such as one would rarely find anywhere except in Papeete. There was Jean Muller, our literary friend from Switzerland, whose camera is responsible for many of the illustrations in this book, there were several other men of arts and letters, there were six or seven feet and fifteen or twenty stone of Rivnac, there was the kindly pastor Vernier, accompanied by five of his muscular brown parishioners to give me a hand as necessary; there was our sixty-five year old Norwegian friend Skare, whose birthplace was in Minnesota and who had never seen Norway but who nevertheless spoke Norwegian as fluently as any of us; there was Captain Larsen, skipper of the trading schooner *Hinano,* he had come out with Robert Louis Stevenson, forty-two years ago; there was Cridland, owner of a cutter and an island in the Tuamotus, and last but not least there was our particular chum Gosta Bengston, who had been with us almost daily since he had arrived from Marotea, away south in the Low Archipelago, where he had spent three years—alone with his kanakas—planting coco-nuts, and who was presently to leave for the Marquesas to plant some more. An observer remarked that no fewer than fifteen nationalities were represented in that gathering.

Farewelling on such a grand scale naturally occupied some time, but when we had gone the rounds, I found that the clearance had not yet arrived, whereupon we set to and recommenced our adieu. It became a lengthy performance after all. Presently Rivnac begged us to excuse him. He must leave, he explained to supervise the feeding of his guests. 'What! Lunchtime already, and yet no clearance!'—That *Capitaine du Port* was taking liberties!

In the end it was not the clearance that arrived, but two more bills, which I promptly refused to pay. In return the port captain declined to clear my ship. Thereupon I told him roundly that it made no difference to me; clearance or no I would sail whether he liked it or not. I would depart right on the moment, and I would like to see what he would do about it.

Some of my friends tried to intervene and offered to lend me what money I might require, but I was stubbornly resolved to 'show him'.

I fully realised that this was an act of gross international disobedience, bordering on piracy, but at the same time, seeing *Teddy* tied up to the warship, the obvious humour of the thing struck me. At the back of my mind I had visions of that warship, sent in pursuit of a fleeing *Teddy,* firing a shot across our bows in stern command to heave to and surrender, and of myself, camera in hand, taking snapshots to immortalise different phases of the great occasion.

However, we left unmolested, and we were not pursued. Perhaps those bills were of the kind which do not stand too close a scrutiny.

In any case, I had proved that I was a proper pirate.

18

A Race with Water and Death

On the occasion of this memorable departure the wind blew westerly in the harbour, as is often the case when the trade wind is strong from the south-east outside. The westerly wind inshore is really an eddy in the aerial current forming under the lee of the high mountains. Thus at the outset we had to beat against a headwind, which compelled us to make several tacks before we cleared the inner reef and could lay the passage.

In the meantime the wind had been gradually abating, dying out entirely as we were approaching the pass, on either side of which a tumultuous sea broke heavily over the reef. Having lost steerage-way we were helplessly at the mercy of the currents, whichever way they cared to take us. Fortunately, fed by the enormous masses of water breaking into the lagoon, the current raced swiftly seaward, whirling *Teddy* through the gap in a breathless five minutes. I heaved a profound sigh of relief when the reef was cleared and the thunder of the huge breakers no longer sounded so threateningly near.

Yet on our little *Teddy* conditions were none too pleasant. A calm belt lay between us and the trade wind which, sweeping down from over the mountains and promontories of Tahiti, struck the sea level in a sharply defined line half a mile to seaward—half a mile of tumbling tossing seas, which took nearly an hour to traverse, assisted, as we were, only by occasional catspaws from various directions.

However, little by little we approached the edge of the trade wind; but as we drew nearer I fell to wondering if it would not, perhaps, be wiser to shorten canvas. We carried all our usual working kites, mainsail, jib and staysail. By all appearance the trades were blowing with great force, causing a great commotion in the sea.

All at once we were involved in the turmoil. The gale struck us like a blow. For one anxious moment *Teddy* heeled over at a dangerous angle. Lanyards cracked and I almost expected the rigging to go overboard, when, righting herself again, she was off.

Examining the lanyards I found that two strands had carried away at one place of the aftermost starboard lanyard, but that otherwise they seemed in fair condition. A foot of hemp rope lashed along the broken part was all that was required to render them secure. To be sure, the press of canvas was rather severe, but the wind was on the quarter, every inch of sail was drawing, and old *Teddy* simply flew over the waves. My heart swelled with pride at the sight.

Meantime, circumstances would not allow me much time to admire the boat. Presently I discovered that the cabin floor was awash. We were evidently making some water.

Before leaving Tahiti I had noticed that the planks above the water line had shrunk considerably, and accordingly, I was prepared to do some pumping at the start, trusting at the same time that the frequent wettings which the boat would receive at sea would soon cause the planks to swell and render our ship tight again. However, I had not taken into account the possibility of striking it so unconscionably rough at the very start.

I pumped. Pumping our *Teddy* is at any time hard work, particularly hard when the boat is rolling and plunging in a turbulent sea. The pump plunger is a simple device, its principle consisting of a heavy piece of leather at the end of a stick, but very efficacious, giving at least a gallon of water at each stroke. Yet my pumping did not seem to avail. The cabin floor was still submerged.

Having exhausted myself, I went down into the cabin to survey the situation. Imagine my consternation, when I discovered that even while I had been pumping the water had risen three or four inches above the floor.

I ran into the forepeak, looking for the leakage. There, to my horror, daylight showed through every seam, and whenever the boat dived into a sea, the water shot in from both sides, transforming the place into a very shower bath. This was decidedly worse than I had expected.

Back I rushed to the pump. My fatigue was forgotten. I was pumping for our lives now, up and down, up and down. Yet the water crept slowly upwards, inch by inch. My wife strove hard at the tiller with Tony beside her in the cockpit. They depended on me for their safety. I pumped faster. The surroundings danced before my eyes; every muscle in my body was aching, but pump I must.

Then the pump carried away, and I sank down on the cabin top, defeated. Below the water was washing over the first step of the companion. I realised that the deeper *Teddy* came into the water, the more of her open seams would be permanently submerged and the faster she would sink. What could I do? I had no means of repairing the pump.

Teddy, as though she sensed the peril, sped gallantly on, doing surely ten knots. The boisterous seas she pushed impatiently aside. Moorea loomed ahead. This was a race against time. We had to make that island, or drown.

I looked below and was horrified; the water had risen over the benches on the leeward side.

Although painfully aware that we already carried rather more canvas than was safe, in my desperation I hauled the trysail out of the sail bunk and set it spinnaker-like to windward. It was a frantic struggle, but I succeeded.

How that sail pulled! Onward *Teddy* flew like a bolting racehorse. I had not the time to log her speed, but I am convinced that she did eleven knots or more. Oh, she was grand!

In the cabin the water splashed about, wetting and ruining one thing after another. It rose inch by inch, a surging mass of rusty water in which at times became visible cushions, books, pulpy biscuits, saucepans, clothing—all the unpretentious things that formed part of our daily life. I noticed them indifferently; greater things were at stake, the lives of my wife, my boy and my boat.

The dog, wet and miserable, cowered forlornly in the companion. She whined pitifully, whenever I looked at her, but even Spare Provisions was of minor consideration then. Presently she came on deck; the water had risen above the companion floor.

I could not keep my eye off that terrible water rising stealthily and inevitably down below. Again and again I would look at my watch and at the island looming ahead; again and again I would try to calculate our chances, swinging between hope and black despair.

The water had established a connection between the sail locker and the cabin; it rushed in and out with the motion of the boat, pounding against the bulkhead in the companion and throwing spray on deck. It had formed into an additional danger, increasing with every moment; I feared lest the wash of water we carried should impair our speed. We could not afford to slow down.

Long since the leeward bunks were soaked. Presently the mattresses were afloat. The water crept up the sides, splashing against the deck. It rose. It rose.

But already the nearest point of the island was abreast. My hopes rose as I edged in towards the reef. However, soon I changed our course again, striking straight for the entrance to Papetoai Bay. The roar of the breakers warned me how utterly futile would have been an attempt to swim ashore. As yet the huge and shifting weight of water inside did not seem to affect the speed of my boat. She rushed onward through blurred visions of sizzling foam, swaying with a deliberate gravity under the heavy burden of her canvas. As the shore sped past, my confidence gradually returned. Only three or four miles more and we would be saved. We would make it, if only the wind did not abate.

A fierce squall struck us. *Teddy* travelled all the faster for it. I could not help laughing. The strength of my boat occurred to me as something immensely funny.

I am afraid that the new-born conviction of our safety had put me a bit off my balance. Already the pass was plainly visible.

Off the entrance I had a hard struggle, when the trysail had to be stowed. Then, after a heavy jibe, we shot through the foam-bordered reef passage into the smooth haven of Papetoai Bay.

A native in a pirogue was fishing in the lagoon. We picked him up on the way and secured his pilotage to a beautiful little cove near the head of the bay. The anchor was dropped, a stern rope passed ashore and made fast to a cocos palm. I looked at the watch: we had travelled more than twenty miles in one and three-quarter hours. Nor had there been any time to spare. *Teddy* lay very low by the head. In the forepeak the water reached up to my chest, and even in the cabin the surface was on a level with the table. One half-hour longer at sea would have finished us.

Now the danger was over. No longer exposed to the continual duckings of the sea, the boat drew water only through the seams which were submerged by the surplus weight of the flood inside, although surely that was bad enough, and we lost no time in starting to bale out the water. Terii, the native, assisted in our toil. Passed hand to hand, three large buckets were permanently in action for three and a half hours, before the water level was at last brought down to the cabin floor; by rough estimate I calculated that we had baled out more than twenty tons of water; at the outset we must have been dangerously close to the point where our combined efforts at the baling would have been insufficient to keep the boat afloat.

We slept wonderfully well that night in spite of the discomfort of our soaked bunks. For Tony we had somehow found a dry spot.

Curiously enough it so chanced that the day we left Papeete marked the anniversary of our eventful departure from Curaçao. In future I shall eye that historic date with suspicion.

When on the following day I surveyed the boat, I found that the topsides had shrunk all the way down to within an inch or two off the waterline. Now that the paint which had covered the seams had been washed away, I discovered places where I could almost push a lead pencil through the seams. No wonder she was leaking!

I think that particular experience has taught me a lesson.

19

Moorea, Raiatea, and Bora-Bora
Three Pearls of the South Sea

Repairing the havoc wrought in our equipment and restoring order kept us busy for several days. Fortunately it so happened that our anchorage was situated within a stone's throw of the bungalow of an American planter, Mr. Kellum, who readily placed at our disposal the various conveniences of his place. On the lawn of his garden we washed out the salt water from pillows, cushions, mattresses, blankets, clothing, charts, photos and whatever else could be saved from ruin by an abundance of fresh water and by subsequent exposure to wind and sunshine.

The water, we discovered, had effected a most thorough-going displacement of articles. Thus the coffee kettle was only found after an ardent search below the flooring in the bilges forward, whereas the lid belonging to the kettle had been deposited on a shelf in the aftermost bunk close under the deck.

Although, naturally, many things were destroyed, some of which were of considerable value to me, yet, once the work of cleaning and drying and general restoration had been accomplished, it was hard to credit how extensive had been the damage.

But for months after, whenever I used the pump, the bilge water would bring to light fragments of manuscript, of poems and other rubbish I had been hoarding for years, lacking the courage either to publish the stuff or throw it away. Fate, it seemed, had taken a hand, and I felt, as I recognised some familiar fragment of poetry in the bilge water, that, on this occasion at least, the gods had done me a good turn.

Moorea is surely the most beautiful island in the seven seas. Particularly fascinating were the nights when the trade wind sank into silence, when the perfectly polished surface of the bay reflected in minute detail the charms of the tropical landscape around. Above the forest-clad slopes beyond the plain rose the rugged peaks, steep and forbidding, peaks which were once, according to legend, haunts of the lizard men, who were ever at war with the people of the plains.

There were not many white men in Moorea, but the natives were very friendly to us. Particularly so was Terii, who every day brought us fruit and vegetables.

Although I was tempted to linger, I realised the advisability of crossing the western Pacific before the approaching hurricane season rendered conditions too unsafe; thus five days later we were once more ready to get underway.

The open seams of the topsides I had filled with white lead and finished off with two coats of paint. Mr. Kellum provided me with leather for the pump and replenished our stores with many useful products of his soil.

We next called at the island of Raiatea. Our intention had been to make only a brief stay here, but as the mainsail had been badly torn on the way, we were forced to remain until the damage had been repaired. In Raiatea we met Dr. Rollin again for the third time. He had been transferred to the post of medical commissioner to the western group of the Society Islands, comprising Huahine, Raiatea, Tahaa, Bora-Bora and Maupiti. He introduced us to the administrator of the group, who in his turn gave us a dinner party, a very amusing performance.

With true French politeness everybody endeavoured to include us in the gay conversation. In particular there was a young man of solid frame, who made heroic efforts in this respect. Although we had not the least difficulty in understanding his French, he was definitely convinced that we had, which conviction, however, did not discourage him. Raising his voice he would repeat his remarks on the weather or the quietness of the place, finally shouting at the top of his stentorian voice to the quiet amusement of the other members of the party, whose conversation he drowned. Then leaning forward with knitted brows to understand our replies, he would misconstrue their meaning and roar with laughter until we could not help joining him. A great lad was he.

Moored as we were alongside the wharf, we had hosts of native visitors, who brought us presents of island produce, especially pineapples, which were just then in season. Wonderfully sweet were those Raiatean pineapples.

Often we would watch large pirogues rigged with a lateen sail racing each other across the lagoon. They carried natives from Tahaa, the neighbouring island, which is linked up with Raiatea by the coral reef which encircles the two islands. Sometimes these outrigger canoes would attain a truly marvellous speed, skimming over the surface of the smooth water like hydroplanes, a native permanently stationed on the outrigger arm to balance the boat. From the precautions taken in hanging the luggage high up the mast one could draw one's own conclusions as to the shelter afforded. The distance between Raiatea and Tahaa is some eight miles and the lagoon thus yields a spacious playground for this type of craft, which by virtue of their shallow draft can sail over the large areas of coral that encumber the lagoon and render navigation difficult for vessels of greater draft.

During our stay at Raiatea two native girls had attached themselves to the crew of our boat, coming on board at dawn and staying till bedtime. They proved not only ornate, but also useful, assisting in our domestic toil and proudly taking Tony for promenades. Neither understood a word of French and our conversation therefore was carried on chiefly by signs and gesticulations. I believe that they had made up their minds to join us indefinitely and when we departed, leaving them on the quay, their grief was genuine and undisguised.

It had been agreed that Dr. Rollin should come with us to the island of Bora-Bora, whence reports had come of a case of suspected leprosy.

We left Uturoa—the village at Raiatea—at ten on a Sunday morning a week after our arrival. The day was bright and sunny and although the breeze was rather light we expected to make Vaitape, the principle village of Bora-Bora, before dark, the distance being only about thirty-five miles.

The first part of our journey carried us across the lagoon; a novel and pleasant experience. The bright sunlight made the channel easy to recognise and yet left sufficient scope for care in avoiding coral heads and shallows to render even the navigatory part of the trip interesting.

Having passed the south-west point of Tahaa, we duly found the pass leading into the open and thence struck straight for Bora-Bora.

Like the rest of these islands, Bora-Bora is surrounded by a coral reef, which has, however, only a single narrow break on the western side leading into the spacious lagoon. We made fairly good time across the stretch of open water separating us from the island, but as we rounded the south-west point and came into the lee of the mountains we found ourselves becalmed, so that by nightfall we had not even sighted the entrance.

However, about eight o'clock a light air sprang up and, although it blew dead against us, I decided to attempt the passage. Guided by the sound of the breakers on the reef we managed to find the opening and then, tacking at short intervals, we groped our way through the pass into the lagoon.

Here again we were becalmed. I sounded the foghorn and lit a flare to attract attention, but my efforts met with no apparent response. The lead gave thirty fathoms of water. Too deep for us to anchor.

There was not the vaguest breath of air. The lagoon lay motionless in the shadow of the gigantic mountain like the glassy surface of a vast sheet of pitch. From a Himene house somewhere on land sounded right prettily the harmony of a song. The shores lay in deep darkness. Once or twice a flickering light would move along the border of the lagoon for a little distance, only to disappear again when we hailed. Our own voices resounded abnormally loud in those enchanted surroundings. At last, from out of the shadows a pirogue approached us. A soft native voice spoke through the darkness.

The doctor called the natives alongside, and despatching them with a written message to the permanent white population of Bora-Bora—the French gendarme—he finally contrived to stir the place into activity.

Presently a very apologetic gendarme arrived with a flotilla of pirogues, two of which, each manned by eight paddlers, took *Teddy* in tow. When, a little later, we moored at the end of the pier, it was about midnight.

The pier proved to be a long elevated roadway built of slabs of coral and metal over the shallow shelf, and ending, where the coral abruptly dropped away, in a wooden structure, which afforded just one berth.

The doctor and the gendarme presently took their leave.

However, the village had been roused and even at that late hour the unusual arrival of a strange ship had attracted a score or more of natives to the wharf, each apparently accompanied by at least one dog, forming a collection of snarling, ill-tempered brutes. While I was making fast my sails by the light of a brilliant moon, rising over the solitary summit of Mount Temanu, I cast frequent apprehensive glances at Spare Provisions, lest she should step ashore.

Quite suddenly she vanished, and instantaneously a terrible row started on the wharf. A dog fight! Some big curs had fallen upon a smaller dog and were apparently tearing her to pieces. From all directions dogs came rushing to the scene and joined in the scrummage. In a few seconds it had developed into a huge heap of kicking and wriggling bodies, whence emerged a terrific medley of snarling, howling, whining and growling. Through it I heard the pitiful voice of Spare Provisions in utter distress.

I jumped to her rescue. Kicking and hitting those curs for all I was worth, I shifted them through the air in all directions. Back they came like boomerangs. Eventually I succeeded in penetrating to the bottom and rescuing the poor dog, and with it in my arms I beat a hasty retreat to the boat.

Then I experienced the surprise of my life: Spare Provisions stood on deck eyeing me with calm disapproval. She had just come up from below to see what all the disturbance was about. Never engaging in dog fights herself, naturally she does not like her master to do so. I looked at the dog in my arms: it was a perfect stranger to me. The natives around were chuckling with glee. Julie sat on the coaming, convulsed with laughter. So, seeing no way of disguising the fact that I had made a fool of myself, I kicked the hairy price of my bravery ashore and went below.

20

Strange Happenings at Bora-Bora

Bora-Bora is an almost circular island surmounted by a central inaccessible mountain, rising steeply to some 2500 feet. It is very picturesque.

The natives are of purer blood than in any of the other islands which we had visited. Having brought letters of introduction from Pastor Vernier to two of the chiefs, one named Tetuanui Marama a Temarii and the other, the tavana tuhaa of Vaitape, Teaotea a Tiavaehaa, we were treated with great friendliness by the entire population. One afternoon, while the doctor and I were fishing off the end of the wharf, I noticed a procession of some fifty men approaching. On poles over their shoulders they carried various burdens.

'Aha!' said the doctor, 'they are bringing you presents.'

He was right. When the men came to the end of the pier, they laid down their burdens in a pile on the ground. I did not know whether to laugh or to cry, when I beheld the 'little gift'. It must have been about a ton of eatables; bananas, coco-nuts, yams, taro, breadfruit, live fowls...

While I bethought myself of something funny to say in order to overcome my embarrassment, one of the party stepped forward and addressed me in his native tongue in such a fashion that I soon realised this was no occasion for frivolous grinning.

Of a surety, I did not understand a word of the speaker's language, and yet it was both elevating and pleasurable to listen to him. His well modulated voice, his expressive gestures and his clever pauses made it obvious to me that here was an orator of no mean ability. When thereafter Hinarai, the interpreter, translated his speech into French, it struck me that Hinarai was merely repeating what I had already understood.

These men wished to honour me and show me kindness, and I must needs prove my appreciation. It proved easier than I had expected. In fact, after listening to a few sentences, I was so impressed by the honest sincerity of these people that it certainly called for no pretence on my part to meet the situation.

I replied with a speech in French which, I fear, lacked all the elegance of form of the first address, but which was, at least, perfectly sincere.

It was indeed a pretty ceremony and one which my wife and I shall always cherish in our memories.

To be sure, this superabundance of food caused me a great deal of trouble before I succeeded in stowing it away in a satisfactory manner. There were twenty bunches—half a ton—of bananas, which it seemed almost impossible to place until at last I hung them, ten each, on two solid poles lashed securely to the shrouds, one on either side.

Afterwards, at sea, these bananas proved a permanent danger, chaffing against each other with the motion of the ship and dropping all over the deck to make traps for the big feet of a sailor hurrying forward at night to attend to some necessary detail of navigation.

About thirty miles north of Bora-Bora lies a solitary and uninhabited atoll, called Motu Iti, a ring of coral without a single break through which to enter the lagoon. To this island the natives from Bora-Bora go to catch turtles, which abound around its shores.

A native returning from a trip to Motu Iti presented us with a big palm leaf basket full of turtle eggs, which we found tasty enough at first, but so very satisfying that, after a few experiences, the greater part of them disappeared into the interior of Spare Provisions.

In the shallow water on the side of the pier lay a turtle shell fully five feet in length, probably left there for the small fish to clean out the remnants of meat. It was always crowded with a multitude of gaily coloured fish, which I often found pleasure in watching.

One day while I was thus idly engaged, suddenly from out between the piles of the wharf protruded the head of a conger eel, a veritable sea monster, if ever there was one. As it warily reached out towards the turtle shell, I beheld about four feet of a neck which surely must have measured at least ten inches in diameter. It was a most uncanny sight, which made me shudder even as I sat there on the stringer. The next instant I rushed on board to get my spear. That eel offered a safe target. I could not miss it at such distance. The monster would probably ruin my spear, but what a trophy to show the world! Then let anyone deny the existence of the Sea Serpent!

When I returned with my four-pronged spear, the eel was still there, but showing only about one foot of its head and neck. I threw the spear with all my force.—Alas! In my excitement I missed and so lost my great chance for ever.

Judging by proportions, that eel may have been between thirty and fifty feet long. I leave it to the zoologists to verify this estimate, but that, at least, was the impression I received, when I saw the monstrous fore part of my sea serpent. Nor do I actually know whether the eel was actually preying upon the small fish, or merely attracted by the very ripe condition of that turtle shell. The latter, I imagine, is more likely.

However, the incident completely spoiled my pleasure in swimming about the pier.

While at Bora-Bora, we spent a day on a tour to the eastern side of the island to visit the German painter Schlesing and his wife, whom we had previously met in Tahiti. It so happened that the doctor was to go almost to the same place, and thus we had the good fortune to make the excursion 'on government service' in a dilapidated old motor boat, the only specimen of its kind belonging to the island. Although the conveyance made no appeal to my aesthetic emotions, we had a glorious outing, and the passage through the labyrinth of coral formations between the palm-clad motus of the reef and the main island was intensely interesting.

The painter lived with his wife amongst the natives in a little village on the border of the lagoon. They were not at home when we arrived, having, indeed, gone across the island to Vaitape to call on us. Meantime the villagers made us welcome, bringing us refreshments in the shape of green coconuts and heaps of fruit, and, when we prepared to leave, presenting us with rare sea shells as souvenirs. Just then the Schlesings returned, but we could not stay; the summons of the conch shell, which we had agreed on as a signal, sounded ever more insistent and presently a runner arrived to say that we must return immediately so as to make the passage through the corralling intricacies before dark.

The owner of the motor boat was a man of the world. For several years he had been sailoring in trading schooners and had been even to Auckland and San Francisco. Yet he was full of the superstitions of his people. Towards dark, when we passed Farepiti Point, in a mood of strange animation he told me about the old 'marae' on the point, which was the haunt of an evil spirit, a 'tupupae'—he shivered as he pronounced the word—which appeared at night in the shape of a savage dog. Remembering my own experience on the first night in Bora-Bora, I was quite ready to agree with him.

However, he was serious, and his very sincerity rendered his talk interesting enough to listen to. He was apparently thoroughly versed in the superstitious tales and legends of his island, which have a strong grip upon the minds of the natives of Bora-Bora to this very day.

Just inside the wharf where we are moored there is a group of rocks amid some giant trees. The natives shun this place. It is tabu: the 'marae' of the kings. Whosoever touches those rocks, so tradition says, will be smitten by the most loathsome of diseases, leprosy, the living death.

Before concluding this chapter I shall relate from memory part of a letter which I gave the doctor to mail to a friend in the Marquesas, a letter which may interest some of my readers:

Bora-Bora, November 1931 (should read 1930)

Dear Gosta,

Doubtlessly you remember that while at Papeete you gave me a piece of material, which you stated to be fragment of a large lump found by you or rather by your fowls on the beach at Marotea. You may also remember that I held it to be ambergris, and that you laughed at me, stating that you had left a hundred weight of

> it for the fowls to feed on. Now I have made certain that it 'is' ambergris and therefore it appears to me that you have held a winner in life's great sweepstake and thrown away your ticket.
>
> It is presumably useless to hope that the ambergris promptly choked or poisoned the whole bunch of those fowls. After all this time you would probably find that the sea, which brought it, has taken it away again, if no one else has. Besides, you said that your poultry throve on the stuff and laid many eggs. Well, they certainly ought to do that at least, considering that those five-score fowls were eating the equivalent of a thousand bushels of wheat or a quarter million eggs daily. I am sure that in time you will come to some conclusion yourself that ambergris is not an economic food for fowls. But, my word! How sweet a breath they must have had at the finish!...

I hoped to rub it in, but fear greatly that when my friend received this letter he would merely laugh. The fellow is offensively indifferent to the power of worldly mammon.

21

The Startling Performances of a Boat Avoiding Reefs on the Way to Samoa, and the Equally Startling Experiences of a Crew Encountering Rocks of Social Etiquette in Pago-Pago

We left Bora-Bora on November 8th. Our Bill of Health was issued for Tonga, but on the way we changed our minds and decided to call at Pago Pago, on the island of Tutuila, American Samoa.

Our voyage to Samoa took us past Maupiti at a cable's distance from the huge breakers which rendered the access to the pass extremely dangerous except in a calm sea. Pastor Vernier in Tahiti had given me an introduction also to the chief at Maupiti, but the necessity for haste prevented me from carrying out my original intention of calling there, and after seeing the entrance I felt no regrets at passing by.

Some 200 miles further west we sailed between the atolls of Scilly and Bellingshausen. By night the proximity of these low islands caused me some uneasiness. An inquisitive native at Bora-Bora having pulled off the spring of my watch, we were without a time-piece and consequently without the means of calculating the distance we had covered.

However, we passed between these atolls without sighting any of them, and from then onwards our path continued clear for a long spell.

Meantime the weather had become misty and squally, and with a high and choppy sea running, the wear on our gear was heavy. The mast again began to work, and there was hardly a rope that was not badly chafed. I felt some anxiety on this score, for the season was so far advanced that bad weather might at any time be expected. As a matter of fact, we experienced on November 24th the tail-end of a hurricane that a day or so before had raged in the Fijian and Tongan Islands.

Rose Island is the easternmost outpost of the Samoan group. It is merely a tiny atoll, a circular coral reef with two little islets rising a few feet above high-water mark. One of these is only a barren sandy cay, but on the other, which is a few hundred yards long, coconut palms grow in great profusion. Naturally it is uninhabited; it would hardly be a safe place in a hurricane.

An island of this order would not be visible from afar. Even in the daytime one would have to keep a sharp look-out to discover it at a distance of five miles. On a dark night its visibility would hardly extend beyond half a mile.

Therefore, when, according to my calculations, we were approaching Rose Island during the night of November 24th I stayed on deck and kept a look-out, ever and anon searching the horizon with my glasses. However, the island did not put in an appearance, and when at last day broke, I submitted the horizon to a final careful search and then went to sleep.

Teddy kept steadily on her course.

When I came on deck again shortly before noon, I sighted the mountainous Manua-group on the starboard bow. Assuming the distance to the nearest island to be thirty miles, I determined our position by bearings. Then calculating backwards on the course which our boat had been travelling, I found that we had sailed straight across Rose Island. And we had not even noticed it! There are surely not many boats that would sail right over an atoll. Truly, *Teddy* is a marvellous craft!

Seriously, however, it must have been a sufficiently close miss. On this occasion I had evidently cut my navigation just a little too fine. But good luck still held.

On November 26th about noon we made Pago-Pago, sweeping into port at a speed of nine knots, having taken eighteen days to cover a distance of about 1200 miles from Bora-Bora.

We were given a most friendly welcome by the officers and men of the naval station. The lack of a clearance and of an American Bill of Health was obligingly overlooked.

The first night the crew of *Teddy* dined with Commander and Mrs

Spore, the skipper in his uniform of shorts and shirt, the other members of the crew attired in their best, albeit somewhat mildewed and locker-creased regimentals. It was a delightful meal to me and Tony but a terrible ordeal to his mother. Although the Commander's own boys sat at a little table by themselves, Tony, in his capacity of chief officer of the *Teddy*, was placed at table with the adults. Though he had never until then handled fork and spoon on his own, he treated these instruments with truly astounding sufficiency if not efficiency. Possibly this was in deference to his company because never since has he equally distinguished himself. Apart from such trifles as ordering fresh supplies before he had finished what he had, he behaved fairly creditably, and as the meal drew towards its end and Tony showed very unmistakable signs of having had enough, the tension on his mother's face gradually relaxed. Alas, her relief was premature! Tony, when he could not eat another spoonful and still saw several dishes at his elbow, promptly emptied them all onto the table and in a second with hurrying hands smeared the contents all over the beautiful table-cloth. Our hosts were gracious enough to think the incident amusing, but I am sure that Julie did not, and even I, his father, cannot but agree with her that Tony's debut in polite society was not a success.

Pago-Pago does great credit to American order and efficiency. With its rows of royal palms, its bungalows, its lawns, tennis courts, golf course and concrete pavings it might well have been a tiny suburb of Los Angeles or Miami. That, of course, is highly desirable from the point of view of the comfort and convenience of the residents, but it is certainly not Samoa. Unfortunately the abominable weather which prevailed during our stay at Tutuila did not encourage the undertaking of excursions into the bush. When one is unfortunate enough to strike the rainy season in an island where the annual downpour averages 250 inches, one will soon discover that excursions into the bush offer few attractions. A resident at Pago-Pago told me that one night he had measured 22 inches of rainfall in ten hours. I cannot testify to the veracity of his statement or the accuracy of his instruments, but in any case it takes a vast amount of rain to make a man set forth such a statement.

Thus, on account of the weather, we did not see much more of Samoa than does the average tourist, which includes the usual show-Indian performances. However, I learned to appreciate, amongst the products of native industry, the remarkable wood carvings and wonderfully woven mats, which are surely not excelled by any other people in the South Seas.

Also I paid a call to the Governor of American Samoa. Seeing that we were being so well cared for by the Navy and that the Governor is the senior naval officer, it was obviously the right thing to do.

But the affair had its difficulties. We had no nickel plated electric irons. Ours were the old-fashioned solid type, which are equipped with half-burnt wooden handles. One of my first duties in every port consisted of chipping rust off the irons and polishing them with sand and emery paper. It usually took hours to render the irons fit for service, but even then they would give the first half dozen things that were ironed a somewhat irregular stove-pipe shine.

However, even these obstacles were overcome and presently I stood forth in a white suit, shining with starch and pearl buttons. My tropical helmet was speckless. My tie was carefully done into an alluring bow. 'The glass of fashion and the mould of form'—save for my feet.

Certainly they were shod in my best, but when at sea I had discovered that the cockroaches were eating holes in the canvas; I had taken these shoes for everyday wear with the result that, after a while, the cockroaches left them alone but my toes worked through the canvas instead.

Accordingly, preparatory to calling on the governor, I went to buy a new pair of canvas shoes at the village shops only to discover that unfortunately none of them stocked my size—a large number ten.

What could I do? I surmised that the governor expected me and I could not go barefooted. A white man 'must' wear shoes in Samoa. It is his indispensable tribute to racial distinction. Seeing no better way, I went back on board and whitened the 'tout ensemble', shoes and toes alike.

From a distance it looked really well, but, of course, at close quarters the effect was somewhat startling. Still there was always the hope that the governor wore spectacles and had opportunely left them at home.

In any case, why should I worry? I had certainly heard that clothes make the man, but I had never yet been informed that shoes have a similar creative efficacy. So off I went, quite pleased with myself.

The governor received me right away, his aide-de-camp made the introduction. Conversation ran pleasantly enough, although I noticed that both gentlemen took pains not to let their gaze fall below my eyes. Personally, I could detect no particularly interesting features about the ceiling. Therefore, in order to relieve their embarrassment and my own uneasiness lest the governor should think that I had not donned my very best shoes for the occasion, I broached the subject myself to the relieved delight of the two Americans.

It proved a good move on my part, resulting, as it did, in a permit to purchase what I needed from the Naval store, and when I took leave, I felt that I had spent not only a pleasant but also a profitable half hour.

His Excellency did me the honour to pay a return call to the *Teddy*, a courtesy which I greatly appreciated, although it came upon me just a trifle unexpectedly.

With my hands and my arms well soaked in a mixture of Stockholm and coal tar, I was engaged in setting up the rigging, when on the wharf, twelve feet above our deck appeared the spotless and dignified white form of the governor.

I offered to fetch the ladder which had just been removed by someone ashore, but his Excellency, being a sailor, would not hear of such a thing, and in spite of my alarmed protests he slid down the well-tarred shrouds onto the still more abundantly tarred lanyards. Well! Well! There was no reason to avoid shaking hands with the governor after that, was there?

22

Piloted by Providence ~ Master-Mariner

Two weeks we stayed in Samoa, re-stepping the mast and overhauling the rigging. Heavy rain and wind prevailed all the time.

About four o'clock on the afternoon of December 10th, we set sail, bound for New Zealand—or somewhere else. What mattered it? The seven seas were ours.

Time enough to decide on our destination, when we had cleared the hurricane zone and crossed the southern trade wind limits!

The weather was rough and as we worked our way in to the open ocean, the wind kept steadily increasing. As we cleared Steps Point, the jib, which had been thoroughly mended in Pago-Pago, carried away again, our usual sign that it is time to shorten sail. I took in the mainsail and set the storm jib and the storm trysail, keeping the full fore staysail on her for speed.

What with bending the sails, donning new running gear and attending to the thousand and one details of preparing for sea, it had been a strenuous day. Therefore, when at length the storm sails were set shipshape, I was well nigh played out, too tired even properly to stow the mainsail. However, the spars had been securely lashed to the lee rail, so that there was no chance of the sail breaking adrift. Passing the gaskets could be left till the morning. I pegged the tiller and brought out a riding light. Then, drenched and exhausted, we all went below. A strenuous day indeed!

On the following morning I found myself incapable of attending to the mainsail; I had contracted influenza and had a high temperature. Besides this, a poisoning in my right hand had begun to worry me. My wife and the boy also were attacked by the flu, but fortunately their cases were light and they soon recovered. This was the first and only occasion on our cruise that any member of the crew had been sick. It was surely a very unpleasant experience, and if we had not been favoured by a continuous chain of lucky circumstance, anything might have happened. To be sure, the initial gale did not last very long, neither did it attain great violence; yet in time I felt very thankful that conditions had compelled me to shorten sail when they did. The canvas of the mainsail remained floating about the deck for three weeks.

As to the sails which we carried, I knew that in an emergency Julie would be able to handle them by herself if a gale should arise; all that would be necessary for her to do would be to lower the staysail. The storm jib and trysail were small sails of exceedingly heavy make and worked by solid running gear. They would be quite safe to carry even in a heavy storm.

Meantime the days passed on, each finding the skipper annoyingly weaker and more feverish. The poisoning had spread up the arm, so that finally it became necessary to split open the shirt sleeve right to the shoulder. The doctor in Pago-Pago had replenished our medicine chest with a variety of useful medicaments, amongst which I found a liberal supply of Epsom salt. This I used freely internally and externally, besides taking large doses of quinine and aspirin to fight the fever. Nothing seemed of much avail. While I was always drenched with perspiration, I was also shivering with icy cold. My coughing shook the boat and scared the dog. Adding to this the ever spreading infection of my arm, I was in a most miserable state. After a while it was all I could do to drag myself out of my bunk to take a noon altitude if the sun was shining.

My condition began to trouble me. It struck me that the almighty owner of the *Teddy* might have it in his mind to discharge the skipper. Therefore I tried to teach Julie to work out a noon latitude, giving her at the same time hints from my rudimentary store of nautical knowledge, pertaining to courses, winds, steamship lines and so forth.

No less important, it seemed to me, though much more unpleasant, was the task of instructing my wife how to dispose of 'the body'. It was evident that if I should die, Julie would not have the strength to lift my heavy carcase out of its bunk and carry it on deck. Yet, if this should happen, she would presently be faced with the immediate necessity of committing it to the waves. It must be remembered that we were in the tropics. But whenever I broached the subject, my wife would ram her fingers in to her ears and flee. Thus it was only gradually and by subtle means that I eventually succeeded in bringing home to her how to bring the jib halyards down into the cabin and hook on, then go on deck and hoist, and finally LET GO!

In this manner I arranged for all possibilities likely to occur, and at last felt content that the two members of the crew would at least stand a fair chance of reaching safety, unless we met disaster before clearing the islands.

The cares and wearisome drudgery of attending to the needs of a sick skipper and of a little boy, washing and airing clothes, preparing meals—all the countless details involved kept my wife toiling incessantly from early dawn till late at night—and these labours on a little boat under such circumstance are infinitely more trying than in a well-ordered household with its many conveniences.

Thus, as there was no one to steer the boat, *Teddy* was left entirely to her own devices.

At the outset, the south-easterly gale had carried us well to the westward, which rather pointed to our taking the course between the Tongan and the Fijian Islands. But as the wind changed to east-north-east the boat chose to head so as to pass the Vavau Group on the eastern side. I did not mind as long as the boat held a steady course, which took her clear of reefs and islands. On this trip my navigation was more than usually sketchy.

For nearly three weeks sails and tiller remained practically untouched. As though guided by a higher power, the boat found her own way among the islands, and even the wind seemed to change so as to accommodate itself to our course.

In spite of the season the sky remained permanently serene; never had we experienced such a long spell of lovely weather.

At noon on Christmas Eve my temperature had risen above 105. Ever on the point of fainting I was hardly able to drag myself on deck to take my daily sun altitude. My subsequent struggle at calculating proved utterly exhausting, and the results were rather vague, giving our position as being some thirty miles to windward of the Minerva Reefs, two extensive and very dangerous barriers of coral, visible only at low tide.

I had turned in again. Julie was washing clothes on deck with Tony keeping her company. Amid the monotonous rumble of the washboard I heard the low murmur of their voices. The dog lay on the floor beneath my bunk whimpering softly. A marlinspike, swinging with the gentle motion of the boat, tinkled with the clear tone of a bell. Each detail in my surroundings I observed with minute distinctness but as if from a great distance and as if it were something strange which had not the slightest connection with me.

Suddenly an exclamation of alarm from on deck roused me from my stupor. Almost at the same instant my wife appeared in the companion in great agitation to announce that we were surrounded by reefs on all sides.

I struggled out of my bunk. The pain in my arm was awful—and as I staggered towards the companion I stumbled and fell heavily upon my sick arm against the steps.

Well! It was not pleasant, but it was the cure. From then on my arm became better. On Christmas Day my temperature for the first time dropped below 103 and thenceforward, in spite of flu which carried on an unequal fight for a few days longer, I made rapid strides towards recovery.

As to the reefs, which threatened on all sides, they presently revealed themselves to be large schools of porpoises playing at a distance.

This was our first Christmas at sea. My wife surprised me with a large tin of tobacco, and I surprised her with a package of her favourite cigarettes. She had certainly bought both, and I had been aware of their existence all along, but that did not lessen our surprise or render the presents less appreciated.

Tony had long since discovered, taken into use, and worn out his

Christmas presents. However, in the desire to make him happy also on this occasion, we bestowed upon him the ship's only time piece, a dollar watch bought in Samoa to replace temporarily my Omega watch, which had been out of commission since we were in Bora-Bora. The substitute having proved itself of a very capricious disposition was worse than useless to us. Therefore, to delight my little boy I removed the regulator and thus transformed it into the world marvel of a watch covering twelve hours in about one minute.

By the time we passed the Kermadecs I was well enough to take charge once more. We did not actually sight the islands, but a southeasterly breeze carried the sulphurous odours of two of these volcanic isles our way, thus giving us an approximate bearing.

On this journey we began to realise the necessity of keeping Tony occupied to prevent him from mischief, which might easily have grave consequences for us all. It was after we had once discovered our fresh water freely running into the bilges from an open tap that this vital necessity was brought home to us.

However, amusing Tony was not a difficult affair. A short fishing line with a bit of white rag tied to the end of it would keep him contentedly fishing all day. A rope fastened to the back of his canvas harness prevented him from falling overboard and thus, standing or sitting by the rail on the quarter, he would display the perfect ardour and patience of a genuine fisherman, all the more commendable since his chances of catching a fish were somewhat faint, seeing that a wise and prudent mother had vetoed the application of a hook to the end of his line.

Meantime there was a rainbow-coloured and friendly dorado who escorted our boat for a fortnight and who would playfully make a tug at Tony's line once in a while to the boundless delight of our fisherman.

The boy was so engrossed with his fishing that often it was only after repeated calls that he would come to the companionway to be untied and taken below for his meals.

On one occasion he refused to budge. 'Hee dat—hee fick' (See that! See the fish!) He called out excitedly. Finally I went to fetch him, and looking over the side, I beheld a huge shark right in the waterline below the quarter. The monster lay turned on his side so as better to view not the fishing line but the fisherman. I hurriedly swept Tony aside and lashed the bayonet onto the boat hook.

My preparations made, I watched the shark again coming to the surface and made ready to strike. But, before I harpooned the offensive monster, I suddenly decided to photograph him.

That saved his life. I obtained the desired snap, but when I once more picked up my harpoon, I found that the shark, obviously sensing danger, had sneaked away. But, for many days thereafter Tony could not forget his great experience with the 'big, big fick.'

A few days after passing the Kermadecs we set all sails, and, benefitting from a strong north-west wind, made good time towards our destination, New Zealand, which we sighted on the eve of January 4th, the 24th day of our voyage. Crossing the date line on this trip we had jumped from the first of January right into the third, cutting out the second of that month.

As in the case of our approach to Trinidad after crossing the Atlantic, so on approaching New Zealand, after sailing about 1750 miles from Samoa, I relied on the vessels which we should probably meet near the coast to give me my correct position before making a landfall. In both cases I was disappointed, and I had to make out my whereabouts as best I could, relying on my usual good fortune. Nor did it fail me. When, on the night of the fourth of January we passed a group of small islands off the New Zealand coast without knowing which they were, suddenly the light of Cape Brett flashed up on the quarter, giving us our position and proving those isles to be the Poor Knights. We were then just within the utmost limits of the range of Cape Brett lighthouse.

On the morning of January 6th 1931, we arrived without misadventure at Auckland, in spite of the fact that I possessed neither charts of New Zealand nor a pilot book to guide me. The lack of charts was due to the unfortunate circumstance that such could not be procured at Samoa, and as to the pilot book, I had been under the erroneous impression that New Zealand was included in one of the Pacific Islands' volumes, all of which I possessed.

Striking from the Poor Knights for Moko Hinau and thence approaching the coast near Kawau at night, we did not have much opportunity of admiring the coastal scenery, which is deservedly famous. In the morning, however, when we were being towed up Rangitoto Channel by the pilot boat, when I had lowered my sails and found leisure at last to look around me at the sunny shores and at the kindly faces surrounding us, I felt that after all Fate and I had chosen wisely when making *Teddy*'s destination the city of Auckland in the Dominion of NEW ZEALAND.

23

The Great Earthquake

New Zealand offered us a heartier welcome than we had anywhere else experienced; it is certainly one of the most hospitable countries in the world. Congratulations, flowers, invitations, and all sorts of courtesies were showered upon our undeserving persons. I was kept busy acknowledging a continuous influx of kindnesses. After a fortnight, however, in response to an invitation from my New Zealand aunt, who owns a sheep farm in Hawke's Bay, the three of us took the service car to Napier, some 280 miles away, leaving the boat and the dog in Auckland, in the good care of some yachtsmen friends.

On the way we stopped at Rotorua, the centre of the wonderful Thermal Region. We had been invited to be the guests of a countryman of ours, old 'Peter', proprietor of the Lake House Hotel at Ohinemutu.

The incessant volcanic activity which is everywhere in evidence in this district, penetrating an obviously thin crust of earth in the form of geysers, boiling mud pools, fumaroles, hot springs and roaring steam holes, is certainly a phenomenon that no visitor to New Zealand should miss; yet I would not like to live quite so near to the internal furnaces of the earth; the word 'internal' seems to be too closely associated with 'infernal' in this neighbourhood. There is something uncanny about the thought of being boiled alive, even if the process be carried out in the most beautiful surroundings, and therefore, after a few day's stay, we proceeded to Hawke's Bay.

My aunt's farm is situated in the neighbourhood of Hastings, in pleasant hilly country that carries as many as six sheep to the acre. Hawke's Bay, however, is a district that occasionally suffers from long droughts, and at the time of our arrival the country had been without rain for seven months. It did not look at its best. The paddocks were brown and scorched and to me it was astonishing that the sheep could keep in such good condition. However, gradually the farmers around were compelled to send away stock to graze in other parts of the country, and even my aunt found it necessary to do likewise, although her place was better off in regard to water than most of the others.

Consequently there was much shifting of stock to be done, at which work I naturally lent a hand, though, perhaps, I was somewhat raw at this game. Anyhow, I found great enjoyment in riding up and down the hillsides, chasing sheep and cattle, and therefore took my share of it, whenever opportunity arose. My wife and the boy in the meantime, spend happy days about the station, where all were doing their best to spoil them.

In this manner two pleasant weeks had passed when the terrible earthquake of February 3rd, 1931, put a sudden end to all our gaiety. This catastrophe was the worst that has ever visited New Zealand.

Naturally, I can give no personal description of the horrors which passed over the neighbouring towns of Napier and Hastings, when large buildings collapsed like houses of cards, when a hillside, crumbling up, shed its millions of tons of limestone and debris from cottages over the promenade, burying trains, trucks, cars and pedestrians, when huge cracks opened in the ground and the bottom of the sea rose so as to leave the inner harbour of Napier dry. For days the glare of fires, which subsequently broke out in this town, coloured the sky a bloody red.

I can only give an account of my own experience, as I watched the course of events on my aunt's farm, where the 'quake was no less severe.

It happened on a bright, sunny morning, when certainly no one thought of death and disaster.

Sitting on the veranda overlooking a garden aglow with flowers, we enjoyed what in New Zealand is an established custom, our forenoon tea.

Suddenly was heard a growing rumble and almost simultaneously a gradually increasing tremor set the china rattling. 'It is only an earthquake,' said Auntie, trying to save her china, but she had scarcely spoken when we were all on our feet clinging to the veranda rail, while overturned furniture and broken china clattered on the floor. Fierce shocks of ever-increasing violence were accompanied by a thunderous noise and the roar of distant explosions. We were thrown against each other. The house swayed hither and thither; from within was heard through the general crash the terrific rumble of the chimneys collapsing.

'For God's sake, save the boy!' cried Julie. I scrambled to my feet and rushed into the house. The atmosphere was dense with dust from fallen brickwork. Through the fog I saw heavy pieces of furniture tumbling madly about. Doors were wrenched off their hinges, frames split. Mirrors and pictures were dancing on the walls, sending missiles of glass through the air. The noise of crashing furniture and crumbling brickwork, of splintering timbers and breaking crockery was truly deafening. The unruly planks under my feet moved so rapidly that my feet seemed to vibrate like drumsticks, landing unexpectedly far out or athwart each other. Yet I knew that I could hold my balance and I sensed a feeling of deep gratitude towards old *Teddy* who had taught me to do so.

Tony's bed had slid into the middle of the room. He was sitting up with a startled look on his face. I grabbed him and hurried out, just in time to escape a huge hallstand which crashed down and blocked the door behind me.

We all gathered on the lawn. Telegraph poles, trees and buildings swayed drunkenly about, cracking and groaning the while. Deep fissures opened in the ground all around. Suddenly a flood of water, six inches high, gushed over the place. 'Thank God, the water is cold,' I inwardly reflected. At that stage I should not have been surprised to see a crater exploding the crust of earth somewhere near and bursting into eruption.

But presently the shocks subsided, only a light tremor remained for some time. How long the first violent 'quake lasted I have no idea. My watch had stopped at 10.47. The papers, although somewhat vague on the point, seemed to agree that the duration should be counted by seconds rather than by minutes. If that is so, at least I know that they were a succession of remarkably long seconds, fateful to the fortunes and destinies of some 50,000 people.

Aunt Hilda's beautiful place these seconds left in a state of ruin. So violent had been the earthquake, that in every room, on every wall the paper had been torn diagonally into small patches. In the whole house only one picture remained hanging and this, after the disaster, hung face to the wall. The carpeted rooms were littered with bricks, mortar, broken furniture, glass, china, flower pots, while above all was spread a thick layer of white dust.

What remained of the water plant was a mass of crumpled concrete and mud and some twisted and broken pipelines. Folds and overlappings had formed in the paddocks. Cracks had opened everywhere and the hillsides were marked with the devastations of slides, at the foot of which lay dead sheep and cattle smelling abominably in the hot sun.

Shocks occurred with gradually lengthening intervals for months after the catastrophe, but fortunately none of them attained the violence of the first disastrous 'quake.

Although, thanks to the fact that the buildings were constructed of wood, no one on the farm suffered injury, even Aunt Hilda's household had paid its toll to grim death. Her little nephew, who had left the night before the earthquake to resume his schooling at Hastings, was buried under the falling walls of a big department store in that town.

It had been our intention to leave in a few days, but, seeing that I could be of service where every hand was needed, I remained for three weeks longer. Then I returned to Auckland to make my boat once more ready for sea. We had made up our minds to go to Sydney, Australia.

24

Preliminaries to a Race

In common with everybody else, I felt deep sympathy with the sufferers from the earthquake, and I could not help admiring the grit with which they instantly determined to 'carry on and rebuild' amidst the smoking ruins. I was eager to help. My pockets, certainly, were empty, but, seeing that I had to go back to Auckland, why not give a lecture and turn over the proceeds to the Relief Fund.

I went to Auckland. The Royal New Zealand Yacht Squadron seemed to afford me likely victims; they would furnish an audience with whom I had in common a love of boats and blue water, and they would supply the atmosphere for good-fellowship, which, prevailing in that fraternity, would put me at my ease, so that I might even do justice to my narrative. All of which proved correct.

When the evening arrived I found that instead of paying a penalty for entertaining altruistic motives, I actually enjoyed myself in speaking to an interested audience. In fact, it is highly probable that in my elation time would have been forgotten had not my eloquence been checked by pseudo-masonic signs from friend Charley in the background, intimating that the time had arrived for refreshments.

Having already met a fair number of the Squadron members, on this night I was introduced to only about a hundred members I did not yet know. However, my memory for names was not up to this moderate test even, the only surname I was able to remember in the whirl being MacCallum. Thus, after trying for a while at Mr. Err... Excuse me and Mr., ah... O'Darnitt, I gave it up and called everybody Bob and Mick and Mac or John or Jim or George indiscriminately. No one seemed to be particularly offended.

The Royal New Zealand Yacht Squadron and the acquaintances I made amongst its members proved exceedingly helpful to me during my long stay in the Dominion. In the Squadron also I was lucky enough to meet Mick who presently became my inseparable friend and henchman.

Mick, being a draftsman when he was not yachting, assisted me in preparing the chart sketches which were to illustrate my lecture. He proved such a pleasant fellow and he had such a grip of the essentials of each job, that gradually it became the most natural thing for me to do, whenever I had a difficulty to tackle, to enlist the aid of ever cheery and obliging Mick.

He was a permanent member of the crew of a little keeler, bearing some heathenish name, and he promptly enrolled me as a member of the same crew.

New Zealand yachtsmen do not regard their sport as an excuse for wearing fine uniforms with brass buttons, while letting a hired crew handle their boats. Hired crews are practically unknown on yachts in the Dominion, the crews of the boats being composed of members of the yachting fraternity, who thus experience to the full all the thrills, the joys and the hardships of the game. The principle is a good one, making for democracy and sportsmanship, and I believe that it is to some extent responsible for the vast popularity of this sport amongst all classes in Auckland.

I may add that although I have seen finer yachts than those of Auckland I have never seen finer yachtsmen or as events proved stauncher friends.

When I was preparing my boat for sea, it so happened that Mr. Bennell from Melbourne arrived in his 42ft. auxiliary ketch '*Oimara*', loudly challenging New Zealand yachtsmen for a race across the Tasman Sea.

Personally I was not greatly impressed by the '*Oimara*' and when, no response to the challenge being forthcoming, Mr. Bennell's regrets on the absence of suitable boats seemed to become just a trifle insinuating, I offered to ship a crew of New Zealand yachtsmen aboard the *Teddy* and race her against the Australian boat to Sydney, my next destination.

The arrangements I left entirely to the Yacht Squadron. Expressing my willingness to accept whatever conditions were imposed, I was confident that within the limits of their power the Squadron would see to the fairness of the terms.

At the outset the whole affair seemed a trifling one, merely a private race between the *Oimara* and the *Teddy*, but the intense interest shown by the public gave the proposed race an ever-increasing importance.

Certain technical difficulties had first to be overcome. As *Teddy* lacked an engine and as the race was originally intended for craft fitted with auxiliary power, an adjustment of the rules was necessary to include my type of boat. At one of the meetings held in the Squadron in connection with these arrangements, Mr. Bennell more than hinted that there would be a pleasant surprise for me 'at the other end.' I was not a little amused by the tone and manner of this remark, which seemed to convey the speaker's intention to donate a money premium for the 'losing' boat.

The conditions agreed upon were, in short, these: the boats were to sail under sealed handicaps to be declared in accordance with Bermudan rules by a committee in Melbourne and to be communicated to the Royal Sydney Yacht Squadron, who would take the times.

The race was to start from Auckland and to finish between Sydney Heads, a direct course of some 1280 miles.

Oimara was free to use her engine and each boat was at liberty to carry as many extra sails as conditions permitted in addition to the duly measured area of her working sails.

When these conditions became known, the public seemed to believe that *Teddy* had no chance of winning and that it was, therefore, very sporty on my part to accept the challenge, which in actual fact it was not.

I knew perfectly well that in an ocean race my boat would be hard to beat by any craft of her size. The fact that *Oimara* possessed an engine would not trouble me provided we had a decent breeze. Nor did I make a secret of my conviction. Rather I am afraid that in extolling the virtues of my *Teddy* I became somewhat of a braggart, but all in vain, for whatever I claimed in this respect was construed as merely a gallant effort to gloss over the shortcomings of my beloved boat, in no wise to be taken seriously. Even the day before our departure the papers said that the *Oimara* was likely to show *Teddy* a clean pair of heels.

In actual verity, the contest resolved itself into an affair of chance rather than an honest race. Should we be becalmed, the *Oimara*, stocked up with a goodly supply of oil of various grades, would gaily chugg-chugg along on her course, utilizing whatever patches of wind she could catch on the way, while *Teddy* must necessarily stay on the spot until a charitable breeze untied her wings and relieved her from the spell.

If, on the other hand, a wind blew, I felt that we would certainly outdistance the Australian boat; while, given a stiff breeze of fair wind, I was confident that *Teddy* would simply lose the *Oimara*, engine and all.

So far there was nothing in the race for the competitors except the sport and a great deal of expense. With my usual optimism I had tackled the problem without considering the financial element, but presently I discovered that a race would involve an expenditure vastly above the limits of my ordinary exchequer. However, by the assistance of a countryman of mine, who offered to give me or, if I preferred the term, to lend me what money I might require, by a donation from an owner of a newspaper, and by the generosity of the Squadron, who opened an account for me with a local firm dealing in ship's equipment, the financial difficulties were overcome. Furthermore, various members of the Squadron contributed liberally to our stores and presented me with several useful light sails.

The days immediately preceding the Trans-Tasman Race were busy and happy. I met with support and encouragement wherever I turned. *Teddy* was on every lip. The papers had almost a column daily about the prospects of the race. Continually, people would stop me in the streets, friends and strangers alike, grasp my hand warmly and hurry through a monologue somewhat to this effect: 'I know that you are busy and that I should not detain you, but I just MUST shake hands with you and wish you good luck.' And many of them would add guiltily and under their breath, as if it were hardly right to entertain such statements: 'I hope you will win!'

Although all this excitement struck me as being wholly out of proportion to the comparative unimportance of the affair, yet I enjoyed it immensely and I could not help liking the New Zealanders all the better for their enthusiasm.

To be sure, in the meantime two incidents had occurred which made the race more deserving of public interest. The Akarana Yacht Club had donated a perpetual challenge trophy in the shape of the Trans-Tasman Cup, and a third boat had been entered for the race, the Auckland keeler *Rangi*. The appearance of the *Rangi* on the scene gave rise to somewhat heated discussions as to her suitability and seaworthiness. Personally I did not consider her an able boat, but the more to be admired was the spirit which led to her entry.

Rangi was to be navigated by Lieut. Commander Juler, a former member of the *Oimara*'s crew. *Oimara* was skippered by Captain Simons, a fine stamp of a man and a true sailor; *Teddy* I would command myself.

As it proved impossible for any of my yachting friends to come away on a trip of uncertain duration, I was compelled to pick my crew from amongst a host of eager applicants, none of whom I knew. I had at last decided on three lads and had turned down a few dozen others, when Brownie came along and with his smiling face managed to secure an extra berth for himself in spite of the fact that *Teddy* had sleeping accommodation for four only. Nor did I ever regret that I yielded to this impulse. Brownie was just the type of mate I wanted, cheerful, ever-ready, a genuine yachtsman. Not that I was unfortunate in my previous selections; considering the circumstances I had sufficient cause to congratulate myself on picking at hazard a crew so keen, but Brownie was the only one who had experience in boats of a size somewhat similar to *Teddy*.

The start of the race was timed for Saturday afternoon at two o'clock. About noon everything was ready. My wife and Tony went ashore and the new crew had taken charge.

In contrast to my competitors, I had not had the time to clean *Teddy's* bottom for two-months' growth, but as I had previously repeatedly experienced that a change of water generally caused parasites to drop off the bottom, I counted on the blue salt waters of the Tasman Sea ridding *Teddy* of the harbour slime, which accordingly would form a handicap only at the beginning of the race.

25

The First Trans-Tasman Yacht Race

By the time we had hoisted our sails, a large crowd of spectators had assembled on the floating crane and on the surrounding wharves to watch us get away.

The crew were hilarious, glowing with pride and excitement, eager to make good. Two of them had been offered ten pounds in cash by way of compensation for giving up their berths on the *Teddy* to other candidates, but they had disdainfully refused to consider the idea.

The start of the race was to take place at 2 pm. Therefore, at 1.30, amidst a clamour of farewells and the cheers of the crowd, we cast off our moorings. A last kiss to Tony and to my wife as the boat glided along the crane. Then, falling away and gradually gaining headway we passed into the sunny harbour.

Rarely has Auckland harbour offered a prettier sight. Scores of white-winged yachts moved about the light-green waters, responding to the gentle breeze with lazy grace; smart motor craft sped hither and thither. The radiant surface was alive with boats, boats of every description, and happy smiling faces. We cruised about amongst them, exchanging greetings and light-hearted jests, until presently the report of a gun, fired from the tower on King's Wharf, warned us that in five minutes we were to start. We were almost on the line, when a second gun announced the start. A moment later we crossed the line and were off on the first ocean race across the Tasman Sea.

When the starting gun sounded, an instant agitation passed over the craft in the harbour. In the twinkling of an eye, as if responding to one single will, the confused crowd of yachts and launches, hitherto spread all over the place and headed in all directions, had drawn together, heading as we did, and forming around us a princely escort of white canvas and enamel paint, a most pleasing sight. Truly Auckland gave us a royal send-off.

Oimara, who for some obscure reason had been towed to the starting point, was four minutes late in crossing the line. Thus at the outset canvas had gained a point on engine power.

The wind was light but freshened a little as we sailed down the harbour towards North Head. If the *Oimara* was ever to 'show *Teddy* a clean pair of heels' under canvas only, now was her chance. We carried our ordinary working canvas, mainsail, staysail and jib, not a great spread for a heavy boat like *Teddy* in a gentle breeze. Yet we held our own with the light craft of our escort. I believe that this start was an eye-opener to many of the spectators who had not realized that it is the part below the water that counts when judging the speed of a boat, and who had never seen the wonderful lower lines of *Teddy*'s hull.

The Australian boat at the outset carried a far bigger spread of canvas than we did, yet it was only when Mr Bennell set the engine going that his boat was able to gain on us. At North Head the *Oimara* had nearly caught up with us and stopped her engine. At once we set our spinnaker, which gave us an approximately equal spread of canvas to that of the other boat which, from now on, rapidly dropped behind.

Rangi, our New Zealand competitor, was not yet in the race. Being on the slipway for cleaning, she had been delayed by the tide, but it had been agreed between us all that her time should be taken and the delay allowed for at the finish.

While we were steadily increasing the distance between us and the *Oimara,* presently we saw *Rangi,* who had started more than half an hour later than we, coming up astern. Under the huge spread of her racing canvas, she travelled like a witch. About 7 o'clock at night she passed us and forthwith disappeared in the murk ahead.

At the time *Oimara* was about four miles behind. However, during the night she apparently ran her engine, the wind being very light, and when morning broke she had vanished from view. Strangely enough, in spite of the extraordinary noisiness of her crude oil engine, she was heard neither by the crew of the *Rangi* nor by us.

Meantime, when *Rangi* had passed us in the dusk, I mustered my resources to see what could be done to increase our speed. True, this light breeze was *Rangi*'s chance, still I did not like to let her get too far ahead of us, if we could help it.

My researches resulted in a notable addition to our spread of canvas. Firstly we rigged an old mizzen, given to me by the owner of an Auckland yawl, as a water sail under the main to catch the spill wind from this sail, and secondly we rigged a discarded spinnaker from a small boat as a water sail under our own spinnaker. Both contrivances proved highly effective, adding a mile to our speed. I often wondered what *Teddy* would be like if she were given a rig enabling her to carry the tremendous spread of canvas which she had the power to bear. She certainly would give even the fastest of racing machines something to defeat.

My New Zealanders were most enthusiastic about the old boat, which naturally pleased me greatly. I did not wish them to have the feeling that they were merely members of the crew of a Norwegian boat. I wanted them to feel that this was a race in which the element of national jealousy could have no part, a race in which the leading spirit should be the love of the game and the love of a good boat. In my mind there had never been room for any doubt but that my shipmates, as they grew acquainted with *Teddy*, would share the love and admiration for her which possessed me. How could they help it?

Aye, truly, my *Teddy* was a boat to gladden the heart of any sailor man!

In this race I wanted the boys to regard *Teddy* as their boat as much as she was mine, the race their race as much as mine, but above all *Teddy*'s own race.

To this end, as much for their own enjoyment of the race, I encouraged them to use their own initiative and judgement to the fullest extent, myself taking the part of adviser and navigator rather than that of the mandatory skipper, without whose bidding nought can be done. The boys responded nobly to my confidence, never relaxing in their endeavour to give the old boat her chance, and I had the gratifying impression that the crew enjoyed every minute of that race.

After tea the first night, I set the watches, assigning Brownie and Parky to the starboard and Bony and Goody-Goody to the port watch. Thus, besides the helmsman, there was always one man standing by to attend to such minor jobs as might occur. The standby was free to go to sleep on the benches in the cabin, but he was expected to slumber with an eye open, so that a soft call from the helmsman would bring him to life without disturbing the watch below. If the watch on deck should prove unable to tackle a task without assistance, I had asked them to call me, a superfluous measure indeed, since any little irregularity in the movements of the boat, or the miscellany of sounds accompanying them, would be sure to arouse me to instant wakefulness.

On the morning of the second day, as we were approaching Cape Brett, we discovered *Rangi* close inshore, slightly ahead of us. She had certainly not gained on us during the night. Thus it seemed that as long as a fair wind permitted us to carry to advantage our novel fancy sails, racing-rigged *Rangi* was unable to draw away from us even in a light breeze.

However, it was not *Rangi* who worried us at the time. The day fell calm and so did the following day. We drifted about on a glassy sea often without even steerage way, while occasional catspaws from the northwest and west carried us slowly northward. All this time we knew that Mr. Bennell would be running his engine to its fullest capacity, striking a bee-line for our common destination under conditions as favourable to him and as adverse to his opponents as if he had ordered them. Meantime, having entered on an absurd race between engine power and canvas, a race in which the fuel oil was provided without stint but not the wind, we had to make the best of unfortunate circumstances, while hoping for a roaring gale to give *Teddy* the upper hand for a change. So far, our hopes were not rewarded. However, about noon on Tuesday a fine breeze of fair wind sprang up and sent us travelling along in good style.

Now we could have struck a straight course for Sydney, but I had made up my mind to sail along the latitude of 32 degrees south as far as Lord Howe Island, and therefore we steered north-west until we gained that latitude.

This route, while adding substantially to the mileage to be covered, seemed to me to offer the better chances for fair winds. As it happened, in this race apparently it gave us nothing but the disadvantage of a longer distance to travel, but that does not prove that I was wrong under average conditions, and I am satisfied that if I were to repeat the journey, I would choose the same route.

Favoured with fair winds and beautiful weather for a succession of days, *Teddy* continued to reel off the miles at an average of 150 per day. Our hopes rose and fell with the rise and fall of the wind. On the whole, though, we felt certain that we were steadily shortening *Oimara*'s lead and outdistancing the *Rangi*. When finally in the neighbourhood of Lord Howe Island, with a steadily increasing wind, we had covered 185 miles in 24 hours; our confidence rose to hilarity at the enticing thought of beating the motor boat even on time, in spite of the handicap of more than two days of calm. We logged nine and a half knots and still the wind was freshening. In all probability we should be in Sydney on Wednesday, eleven days after our departure. Then all of a sudden, while we were ramping along merrily, the wind was gone—we had run into a dead calm.

At first we naturally hoped for the wind to return, but as the weary hours dragged on without the slightest breath rippling the oily waters around, as the hours grew into a day and yet another day our confidence evaporated. On Wednesday morning between two and three we spoke to a steamer, the S.S. *Golden Cloud*. I seized the opportunity to correct my longitude. During the day the breeze returned, but it was too late. By then, we thought, the *Oimara* would be in port. Unacquainted with Bermudan rules, we did not give the handicaps much thought.

However, from a sense of duty to the boat, we continued to sail her to the best of our ability, covering 171 miles the first day and 135 during the next. The wind was again fizzing out. At last, on Friday night we sighted the glare of Sydney ahead. With only light and variable airs it took us all night and the best part of the morning to cover the distance, and when finally we drifted over the finishing line between Sydney Heads about 11 o'clock there was not a breath of wind left.

The pilot steamer came towards us. A passing liner dipped her flag to us. Pleasure craft, displaying gay bunting, came to meet us. What a lot of fuss, we agreed, to make for a losing boat. The pilot boat drew up alongside. Someone hailed us. Brownie answered. I did not catch what was said. But suddenly I saw Brownie jump high into the air and shout Hurrah! Poor fellow! I had often told him that his brain was a bit on the weak side. But then he turned to me with shining eyes and shouted: Don't you hear, Skip, we have won. *Teddy* has won the race, won on handicap with two days to spare. Hurrah for *Teddy*! Hurrah...'

He broke off and dived down into the cabin. I looked below and saw him emptying the remainder of our liquors, whisky, brandy, and rum, into a saucepan. With a cupful of this potent brew for each member of the crew he presently returned on deck proposing to drink to the *Teddy*.

'Skal! Good old *Teddy*.'

As I have often said: That Brownie is a marvellously quick-witted fellow!

26

Ups and Downs

It proved that the *Oimara* had arrived two days ahead of us, but according to the handicaps declared by the Melbourne committee, the Australian boat had to give both of her competitors time, four days to us and three days to *Rangi*. This was hardly fair to the *Rangi* who had been penalised on a sail area, which at sea she could not carry to advantage, but that, of course, was her own look out and in the Tasman Race it was of no consequence as to the final results, for the *Rangi*, having missed Sydney and gone erring down the coast, did not turn up until five days after our arrival. Thus we had beaten the *Oimara* by two days, the *Rangi* by six. By later comparison it proved that our best day's run was fifty miles better than that of any of our competitors.

Considering the fine weather experienced on the trip, I was pleased enough with our win. Still, the results of the race fell very much short of my expectations. I would much rather have had a rough passage and an opportunity to show what my boat could really do.

Allowing for the adverse currents we had covered a mileage of 1575 when we arrived at our destination. To the *Oimara* the considerations which had promoted me to strike out north to look for favourable winds in the latitude of 32 degrees south were a secondary matter. Utilizing the calm weather to run her engine on a straight course for Sydney, she would hardly have covered more than 1350 miles on the run, the exact distance being rather less than 1280 miles. Thus the average daily distances work out exactly alike for the two boats in spite of *Oimara*'s engine and in spite of the fact that we were becalmed or almost becalmed for five days. That the *Oimara* ran her engine to the utmost is proved by the fact that on her arrival at Sydney Heads, she did not even have sufficient fuel left to take her into the harbour. As to the *Rangi*, she followed a route similar to ours.

It is from a feeling of duty to my boat that I make these lengthy explanations in order to correct erroneous opinions, which seem to prevail in some parts. Not having achieved my principal aim to show what my boat could do, the whole race was next to nothing to me, except that it gave me the chance to pin a rose on to my *Teddy*'s old breast.

Remaining in Sydney for ten days, *Teddy*'s crew was most hospitably entertained by the Norwegian Consul General, the two Royal Yachts Clubs and a number of local yachtsmen.

It had been my intention to make the return journey to Auckland by myself and I rather looked forward to it after all the noise in connection with the race. However, yielding to the fervent persuasions of the combined crew, who feared lest the circumstances might be misinterpreted by an uninformed public as desertion on their part, I abandoned my decision and it was arranged that Bony and Goody-Goody should sail back with me on the *Teddy*, while Brownie and Parky, compelled to return to their work, left for Auckland by the mail steamer.

Our homeward journey was a lengthy and tedious one. Beating against persistent gales of easterly winds we were 23 days in reaching the Three Kings, north of Cape Maria van Diemen, and yet another 36 hours before we cast anchor at Maungonui in Doubtless Bay, where a kindly population treated us with exquisite hospitality.

Two days later we left for Auckland. This was the second time on the cruise that I had started out on a Friday and, naturally, things again went wrong. In the afternoon, off the Bay of Islands, we struck a fierce gale from the south-east. Seeing no object in weathering the gale outside, when we could shelter in a snug port, we ran back to beautiful Whangaroa, where we remained until the blow was over. While we were again preparing to leave, a friendly scow, named *Scot*, offered to give us a tow out of the bay. [For the information of the uninitiated I had better state that in New Zealand a 'scow' is a flat-bottomed vessel carrying one, or two, large centreboards, and rigged as a fore- and-aft schooner.] The

Scot possessing a powerful engine, I naturally accepted the kindness, as the high hills surrounding Whangaroa rendered the bay so calm that it is often difficult enough for a sailing vessel to navigate. However, when we passed the Heads, we found the calm prevailing even outside. Therefore, since nothing more was said, I was well pleased to see the obliging *Scot* continue towing us.

Leaving the shelter of the Cavalli Islands, we met a swell, which presently caused the tow-rope to break, but the well-meaning skipper of our tug turned back to give us another line. For once my luck deserted me. The unwieldy scow in the swell refusing to obey the helm, ran into our stern, banging the full weight of her hull on top of our rudder, breaking the gudgeons and dislocating the shoe. Having no headway, we were powerless to avoid the collision.

The scow towed a disabled *Teddy* to Orokawa Bay in the Bay of Islands.

In this sheltered place we beached the boat and temporarily repaired the rudder. During this period my luck seemed to have left me altogether. Mishap followed upon mishap, until finally the crowning misfortune occurred when one morning in a hard blow we parted our anchor chain and a 5-in. mooring rope and drifted ashore just at spring high water.

The surf presently broke off the rudder entirely and gradually dug the boat's keel into the sand, all of which made me feel very sorry for myself.

However, after a few days the weather moderated and we set about refloating her. This proved a difficult task and was only achieved after several futile attempts. In the salvage three of the local launches played an important part in the temporary humiliation of a dismayed skipper, who had to admit that power craft have other *raisons d'etre* besides creating unpleasant noises and foul smells. Still, the hospitality and kindness of the residents and the ready helpfulness shown me by whites and Maoris alike add a share of pleasure even to the memory of this dark period of *Teddy*'s stranding at Orokawa. Goodwin, having to return to his work immediately, was compelled to leave us on our arrival, but Bony was able to stick to me through all my manifold troubles.

About ten days after our misadventure we reshipped our ballast and gear, and, with the rudder repaired in a fashion which had no relation to the craftsmanship originally employed in her constructions, we struck out on the last leg of this trying voyage.

On May 19th, in the thick of a rain-storm, we arrived at Auckland. Officials of the Yacht Clubs and other friends, amongst whom was Brownie, had braved the weather in coming to meet us, and the Harbour Board with never relaxing courtesy sent Captain Wainhouse with the harbour launch to tow us to our usual berth alongside the floating crane.

Then Bony went ashore. He had played his part as a sailor true to tradition even to the necessary detail of continually grumbling about the grub. When I saw him again the next day, however, he confessed to me that he had gained fifteen pounds on the cruise. Thus, apparently, he had not fared so badly after all.

I came home in time to share in the celebrations of Tony's second birthday, and, having on the way home filled my pockets with chocolate for the occasion, I was made very welcome by a practical-minded beneficiary. However, it was not Tony who was uppermost in my mind when I hurried to Devonport to see my family.

While I had been away, my wife had done her part by dutifully adding a new member of the crew to an undermanned boat, a healthy little girl, on whom we later bestowed the New Zealand name of Tui.

Apparently Tui, the baby, was as pleased to see me as was her mother, and although, in spite of my precedents, I am by no means a fanatical advocate of female seafaring, little Tui won the approval of her skipper and the heart of her father from the very start.

Apart for the time which my wife had spent in the excellent care of a very efficient and kind-hearted nurse, she had been living with some Norwegian friends, who, on my return, extended the hospitality of their roomy home even to me.

About this time I came into possession of unaccustomed wealth, originating chiefly from the honorarium for an article in the *National Geographic* magazine. Seeing that besides the necessary repairs to the rudder the old boat was in sad want of a general reconditioning, I decide to make a proper job of it.

Therefore, when the Devonport Yacht Club, of which club Mick was a prominent member, offered to haul up my boat at their boathouse free of charge, I promptly and cheerfully accepted, rented a cosy little house in the neighbourhood, and went to work on the *Teddy*.

The house we rented was not furnished and the rent was rather above the average, but there was a wide lawn, surrounded by hedges, where Tony could be left to play without running into serious mischief, and there was also a workshop with a carpenter's bench. We took a fancy to the place and therefore nothing else would do. The difficulty of furnishing was overcome by the aid of friends and neighbours who lent us all such articles of comfort and utility as we required.

For four months we lived here, while I worked on the boat, stripping her inside and outside, burning off every bit of paint, cleaning, drying, repainting, overhauling everything, renewing keel bolts, stay bolts, and various other iron fastenings, and introducing improvements in the galley, in the cabin, on deck and aloft.

Mick was my untiring helpmate, toiling on the boat every week-end from Saturday noon till Sunday night, and during the week doing what he called 'staff work' which had a wonderfully reducing influence on the item of expenditure. In many cases when Mick asked for a reduced charge, the supplying firms went one better, and there was no charge made at all. Thus, naturally, the money I had at my disposal, went a long way, and when finally we launched *Teddy* again, she was as good as a new boat and more comfortable inside than she had ever been before.

I look back upon those week-ends about the boat with a feeling of gratitude and pleasure. As a rule there were at least four of us working, two young friends of mine, Arty and Fred, being almost as sure to turn up as Mick himself. Yachting and everything connected therewith has a wonderful grip on the youth of Auckland.

27

A Cruise to Tonga

T*eddy* had been equipped with a new set of sails. The heavy gaff, unwieldy in a gale, had been discarded and a leg o' mutton mainsail shipped. This, being made of hemp canvas and of a lesser area than the old gaff sail, was much easier to handle than had been the latter, although, naturally, it did not draw quite so well or drive the boat about in the tacks as lively as the other sail had done. However, as for its suitability for single handed ocean cruising, the new sail constituted a vast improvement.

When everything had been duly tried out cruising about in that yachtsman's paradise which is called the Hauraki Gulf, adjoining the Auckland, or Waitemata Harbour, the proper crew, now consisting of four members, moved on board again and presently *Teddy* departed for Tonga.

We left Auckland on October 24th. With a stiff breeze blowing from the south-west, the weather was chilly, and as we cleared the coast we found conditions boisterous and not too comfortable. Tony, who had been land-lubbering for ten months, became seasick, and although, when asked how he felt, he would heroically reply: 'Feeling brave,' it was obvious that he was dissembling somewhat.

Steering a straight course for Tongatabu, the wind was astern. However, as the sea rose, we were soon compelled to keep the wind on the quarter to avoid the danger of a jibe.

From a nautical viewpoint it would have been correct to hold the wind on the starboard quarter and make easting in these high latitudes. Nevertheless, for the children's sake, I chose to run north, keeping the wind on the port quarter. Thus, after three days, we certainly found ourselves in pleasantly warm weather but slogging against a north-east wind, which would not allow us to lay closer than two points off our course. The passage was rather uneventful. Beating against a trade wind on a 1000-mile course is a slow sort of entertainment, although the children contributed much to make it endurable. Tony, after twenty-four hours of seasickness, had found his bearings again and demonstrated his love of the briny ocean by keeping permanently soaked with it, his chief occupation, besides fishing, consisting of playing with water, which he hauled up from overside by means of his own little pail. That pail was his great treasure. Together with a knitted monkey of indeterminable species it shared his pillow at night.

One fine day, as we were jogging along peacefully at four or five knots with tiller lashed, the handle came off Tony's bucket, when he threw it over the side to haul up water. At first he seemed to be stunned, looking helplessly from the handle which he held in his hand to the brightly-coloured bucket jumping up and down in our wake at a rapidly increasing distance. But then, as the full extent of his loss came home to him, with a pitiful outcry he threw himself face down on the deck and wept, a picture of utter misery.

It did not take us long to discover that we had not the heart to regard this overwhelming grief without at least making an attempt at salvaging the object thus cherished. We came about and retraced our track. The bucket was slowly filling with the splash of the waves. It was a question of reaching it in time. At least there must be no miss. If I did not grab it the first attempt I would have no opportunity of a second. Tony had stopped crying and was watching the proceedings with tense excitement. The recovery of the bucket had become a matter of the utmost importance to us all. I caught sight of Tony's pleading eyes, as I hung outside, wishing fervently that the wash from the bows would not sink the toy. It came through those waves almost full of water and I managed to grab it just as it was preparing for its final plunge.

Tony's joy and the abandon of his gratitude as he hung about my neck afterwards repaid me a hundredfold for the trouble I had taken in fishing up a dilapidated sixpenny pail. From then on for a long time Tony regarded his father with the added affection a child would feel for an amiable and powerful sorcerer.

Two weeks after our departure from Auckland we sighted Pylstaart, or Ato, but the wind was light and a strong westerly set impeded our progress. At dawn on the seventeenth day we were close under the lee of Eua, a mountainous island south-east of Tongatabu, when the trade wind returned with fierce squalls from the north-east, gradually settling down to a moderate gale. We did some hard sailing that morning in order to weather the eastern point of Tongatabu against a sweeping current, and we only just managed to scrape clear with the roaring coral reef barely a cable to leeward. Thence the passage to Nukualofa runs for about ten miles in a westerly direction, winding its way between coral reefs, shallows and islands. Wearing off, we presently had the wind right aft, rolling through the pass in an awkward following sea and with a strong current sweeping us along. Therefore, to avoid a jibe and to give me more time for navigating and for keeping a lookout, I lowered the mainsail, not an easy task under those boisterous conditions. Nor did I have time to finish the job properly. My navigating kept me busy enough. Continuously consulting the chart and the Sailing Directions, spying for coral reefs and keeping account of the little palm-clad islands we passed, one of which was as similar to the other as a string of pearls, I was buzzing from one end of the ship to the other, looking for landmarks, taking bearings and applying them to the chart in the cabin, up and down, scores of times. At the same time I kept an anxious weather eye open for approaching squalls. The weather was still threatening, and a squall, such as we had encountered earlier in the day, would have forced us to heave-to instantly, and even then our chances of escaping disaster would have been but small. Coral reefs are visible only in bright weather. In a rain squall, with the landmarks blotted out and a strong current sweeping to leeward, our position in that labyrinth of coral would have been precarious. It was important to keep track of our progress incessantly.

Yet it was also necessary to change the jibs and set a trysail, for at one place the passage turned at sharp angles to the northward, narrowing in at the same time to the width of a cable and rendered dangerous by a rock lying in mid-channel and by the fact that in the narrows the current sweeps sideways across the fairway. We came through it all unscathed. During the exciting and exhausting hour that it lasted, my wife had been at the tiller, while Tony had been below with strict orders to remain there.

At last we found ourselves at the anchorage off the wharf of Nukualofa. The monthly mail steamer *Tofua* happened to be in port, and the pilot and harbour master (both in one) was apparently engaged on board this ship. We, accordingly, prepared to wait for him before anchoring.

My wife had put the boat head to wind to take the way off her, and the trysail, beating furiously, whipped a heavy sheet block about over the companion hatch. I was forward making the anchors ready, when a startled cry from my wife caused me to look aft. There was Tony, tired of his confinement below, trying to force his way past that frantic sheet block whirling madly about his head. He was in imminent danger: one blow of that flying block would have killed him. Tony, seeing his way barred by that formidable foe, was thoroughly annoyed. "'Top it!' he cried, reaching out with his little hand, "top it!'

Meantime, if the boy was not frightened, his parents most certainly were. I have never had such a scare in my life and I tried to impart some of this feeling to my young son after I had rudely torn him away from his perilous position and deposited him in his bunk below.

Nukualofa is the capital in the little kingdom of Tonga, which boasts a population of 27,000 souls scattered over some 300 islands. It is at present ruled by Queen Salote, whose husband, the able and kindly Prince Tugi, acts as permanent prime minister. The Friendly Islands being a British Protectorate, the majority of the higher officials are Britishers, and it is probably due to the sobering British influence and the judiciously applied authority of the successive British consuls in all matters of finance and expenditure that this little comic opera kingdom, peopled by a careless, happy-go-lucky race of Polynesians, enjoys a financial soundness known in no other part of the world. Tonga has no public debts. On the contrary, the Tongan government maintains a substantial credit balance in British banks.

Thanks to a wise law, which claims all land for public property, preventing its sale and mortgaging, the natives are as undisputed owners of their soil today as they were before the arrival of the white man. At the age of eighteen, each young man is endowed with a piece of land, varying in size according to its fertility and situation. On the fertile island of Tongatabu the allotments today average some thirty-two acres, which is ample to provide for even a large family, although intensive cultivation is rarely seen, for the easy-going Tongans do not bother their heads about the accumulation of wealth. On the other hand, poverty is unknown in these islands.

The little capital is situated on the northern shore of Tongatapu, which in a wide semi-circle to northward is surrounded by reefs and islets, forming a spacious though somewhat coral-infested lagoon, smooth enough in the prevailing easterly winds, but affording doubtful shelter in northerly gales. Like the majority of islands in this group, Tongatapu is of coralline construction, flat throughout except for an elevated edge at the southeastern point, where the land rises to an altitude of 200 feet.

In spite of the profuse vegetation, on a dull day Nukualofan scenery seems rather monotonous. The beauty of the landscape lies mainly in the brilliance of its colours revealed only beneath a serene sky. However, fair weather prevailing almost continuously, Tongatapu, set in an indigo sea with the flash of snowy breakers on bordering coral, with its white beaches, its emerald and sapphire lagoon, in which the reefs show in an infinite variety of transparent hints, with its foliage in all shades of green and its wealth of gorgeous flowers, forms indeed a lovely picture.

Unlike the Society Islands and Samoa, where the bordering coral shelves generally drop away suddenly from two feet of water to abyssal depths, Tonga possesses an abundance of lovely ocean beaches sloping gently into deep water, ideal playgrounds for holiday-makers, were it not for the sharks. Also Tonga has an air of natural cleanliness which the aforementioned islands lack. Whereas in Tahiti, for instance, the scenery lost much of its charm, seen at close quarters, in Tonga it was just the opposite.

There, beneath the palms, naked brown soil full of rubbish and perforated with millions of crab-holes, here mats of beautiful turf. In yon islands it was hard to find a place where one would care to sit down and admire the landscape. In Tonga one might wish to picnic almost everywhere.

In boyhood the appeal of the Friendly Islands would surely have proved irresistible and even to my mature senses they offer a strong fascination.

Captain Anderson was the commander of the Tongan fleet, i.e. the auxiliary schooner *Hifofua* of some eighty tons burden. He was an excellent sailor who had been roaming about in the Pacific for a lifetime. Therefore, when he suggested that I should come with him on a trip round the islands, I readily consented. My wife preferred to take the children and stay ashore, and the *Teddy* was moored to a huge buoy by means of two 6-in. manila ropes, which in Aruba had served to lift the whole boat out of the water. Thus she would be safe enough for a few days.

When the *Hifofua* wound her wake between coral heads and reefs into a thirty-mile stretch of open water separating the southern group from Namuka, her first call, I was on deck helping the Kanaka crew to trim the sails. About five o'clock in the afternoon we dropped anchor off the village and I went ashore with the cargo, some five pounds of meat for the trader, intending to explore the island.

It proved to be a pleasant enough little spot. The western half of the island consists of a ridge of wooded hills, while the greater part of the lowlands in the eastern portion is occupied by a shallow, mangrove-bordered brackwater lagoon. Between the lagoon and the hills lies the village on a grass-matted stretch of land, shaded by sparse palms and giant banyans.

Besides a few hundred Tongans, living in dilapidated huts, scattered about at random, partly by the owners' choice and partly through the activity of last year's hurricane, there were two white traders and an incredible number of incredibly underfed dogs. Bora-Bora faded in comparison to this island. For an abundance of dogs of surprising slenderness, Namuka beats any place I have seen. I passed a remark to this effect to Mr Hansen, one of the traders, who had accompanied me on the stroll, and was surprised to see the grim expression my innocent remark seemed to produce on his face.

'You can hear their bones rattling,' I added jokingly, when a team of dogs trotted by, giving us a wide berth. 'That's not their bones rattling,' replied Mr Hansen savagely, 'that's my copra—inside them.'

I could not help laughing, but hastened to apologize, for Hansen was an interesting old timer in the trade and he had asked me to dine with him.

'Oh,' he said, 'when you know that the only feed these worthless curs get is off your drying platforms, you would wish that they were a sight skinnier. And,' he continued, 'that's not the whole tale. Last night I laid out poison around my backyard to get even with some of them, and this morning I found the yard full of dead bodies.'

'Well—' I ventured.

'My own pigs!' he snapped. 'I believe those dogs must have opened the pen.' He paused. I kept my peace.

'Not a single dead dog!' He presently concluded. 'Except my own.'

The subject of dogs was religiously avoided in our conversation that night. We had chicken for dinner.

Back to New Zealand

Seeing that geographical digressions are beyond the borders of this narrative except in so far as they have some connection with the cruise of the *Teddy,* I will pass over the particulars of this excursion with the *Hifofua,* merely stating that we visited Lifuka in the Haapai group and Vavau, the lovely main island in the northern group, whence we returned by almost the same route. Thus, one fine morning after our departure, we were again on our way to Nukualofa. Namuka lay behind, its white beaches glaring in the sunlight.

Abeam we had Falcon Island, one of the visible outcrops of that long chain of submarine volcanoes which runs parallel to the Tongan Islands right from Pylstaart to the smoking cone of Tofua and beyond. Falcon, now some 200 feet high, has had a remarkable career. It first appeared above sea level some sixty years ago, a mound of scoria and ashes surrounding the crater of an active volcano. Within twelve months it had reached a height of more than 300 feet. Since then it has vanished and reappeared three times, but now it has apparently come to stay.

On my return the crew moved on board again although *Teddy* remained in Nukualofa for another fortnight.

Having promised to be back in New Zealand before Christmas in order to take part in the annual yacht race to Tauranga, I felt obliged to depart not later than December 9th. Thus, when this day arrived, we filled our water tanks, replenished our stores and prepared to leave. Some of the leading white residents, headed by the British Consul, came to bid us farewell, bringing much-appreciated presents in the form of Tapa cloth, Tongan grass skirts and other products of native industry.

It was almost calm when, at noon, we cast off our moorings. However, the current was favourable and, what with the occasional light airs and a spinnaker set, we managed to scrape clear of coral patches and to keep in the fairway until, late in the afternoon, a breeze sprang up which enabled us to make our escape through the narrow western reef passage, before the darkness of approaching night rendered the leading beacons invisible.

An easy swell met us outside. Heading west where the embers of a flaming sundown were still glowing, we presently found ourselves enwrapped by a starlit night of entrancing beauty.

A gentle breeze filled our sails with an even pressure. Rising and sinking with the breathing of an ocean asleep, *Teddy* cut her easy furrow through the sea with a pleasant gliding motion. Softly the water surged by, while myriads of diamonds glittered in our wake.

When the children had been put to sleep Julie joined me in the cockpit. Not a word was spoken between us, and yet, with hearts full of infinite contentment we sat there far into the night, drinking in its beauty. Long since had we passed the coral dangers off the coast. Endless stretches of open water lay before us. There was no need to continue our vigil. I said so to my wife. She did not move nor answer.

Aye, she was right! This was not a night to waste in sleep or idle talk.

At the end of five wonderful days and nights, when the sea, fanned by a light trade wind, remained smooth, and the sky serene, we had covered more than half of the 1200 mile run to Auckland. Then the northeasterly, which had carried us so far, developed into a gale and the gale into a storm. The seas formed into mountains. We were forced to heave-to.

Three days later, the storm having abated, we again set sail and resumed our course, though the sea still ran high. Within twelve hours our progress was stopped once more by a fierce blow from the northwest, causing a dangerous, crosswise sea, which broke over our deck continually and sent cascades of salt water down the companionway. 'It's raining!' said Tony, who, trying to sneak on deck contrary to strict orders, was caught on the ladder by a more than ordinarily powerful waterfall. 'It's raining!' He will be a sailor someday, I predict.

Delayed by these gales it seemed questionable if we should be able to keep our appointment in Auckland. Even when, the day before Christmas Eve, we sighted the Poor Knights, it was doubtful if we should be in time. The wind, on the whole, was capricious, and it was through stubborn resolve rather than through any favours of the weather gods that we finally managed to make the port of Auckland at 2 p.m. on Christmas Eve. I had not closed an eye for sixty strenuous hours.

Mick met us on the wharf. The Tauranga race was to start at 6.30 p.m.

The wind was light and *Teddy*'s bottom was foul with barnacles and other marine curiosities grown to appreciable size during the four months for which the boat had been in the water.

Considering the drawback these circumstances held for the *Teddy* and seeing that we did not arrive in time to clean the bottom before the race, Mick had made no definite arrangements for our participation except for the formal entry. It was obvious that we had not the slightest chance in the race. However, I had come 1200 miles, driven the boat as hard as adverse conditions would permit of in order not to disappoint Mick and the boys, who were only waiting for a telephone call to rush on board and form the crew, and, having so far succeeded in running to schedule, we did not waste much time in discussing our prospects. We had less than four hours in which to make our preparations, which included mustering the new crew, replenishing our stores, effecting various repairs to the rigging and fetching borrowed spars and sails from a shed in Mechanic's Bay. By six o'clock we were once more ready to go to sea, and my family went ashore with some friends, intending to take the service car to Tauranga on the following morning.

As soon as the race had started, I went to sleep. With Mick aboard I had no hesitation in doing so.

No one had expected *Teddy* to shine in the race and no one was disappointed that she did not. The wind remaining light until about noon the next day, *Teddy*'s foul bottom proved too much of a handicap, and when we passed the finishing line of a 130-mile course twenty-five hours after the start, we were the last boat in, the first having arrived some three and a half hours ahead.

Tauranga Harbour Board had sent a launch to tow us up the river to a beautifully illuminated town, which gave us a whole-hearted cheer on our arrival. *Teddy* appeared to be popular in Tauranga. It is a great privilege to have friends in a strange place, and in Tauranga I had a most enthusiastic friend, and a true one, in the person of Jim Denby. Hence the popularity.

We combined the race with a pleasant week's cruise, returning to Auckland after New Year to defend the Trans-Tasman Cup in another race to Sydney, which, however, to my chagrin, did not eventuate. The boat, which had posed as a challenger for more than six months, withdrew immediately before the race, and although we remained in Auckland for two months no other boat turned up to take its place. However, what with cruising about in the Hauraki Gulf at week-ends and enjoying the hospitality of our numerous friends, time passed pleasantly enough.

In the beginning of March, seeing that all efforts to stage a race had proved futile, we prepared for our final departure.

29

Teddy's Last Voyage

The farewells were over.

We had cast off the last mooring rope and shaken the last friendly hand. Slowly, very slowly we glided away from the boat steps. A navy pinnace offered to tow us past the harbour wharves. When, after belaying the tow-rope, I again turned my gaze towards the landing stage, I saw that the crowd on the wharf was thinning out. Perhaps a score of friends still kept on waving: Mick—Jim—Nurse—Brownie—Cobs—Gibbie—Fred...

I suddenly felt a pang of regret, as I realized that I should probably never see again those lovable people who had favoured us with the rare and precious gift of their friendship. Yet, I could not stay. I am not made to live in a well-ordered community for long. I am a wanderer...

> It's like a book, I think, this bloomin' world,
> Which you can read and care for just so long,
> But presently you feel that you will die,
> Unless you get the page you're readin' done,
> An' turn another—likely not so good;
> But what you're after is to turn 'em all.
>
> – *Kipling*

More than thirteen months had elapsed since we first entered the harbour. Many times since then had we sailed in and out, on the Trans-Tasman race, on our cruise to Tonga; many times. On the whole they had been thirteen happy months.

Now Teddy cut her furrow through the waters of the Waitemata for the last time. To starboard arose the well-known cone of Rangitoto, visible from seaward at great distance. To port the sunny splendour of Cheltenham Beach glided by. We tried to pick out the trees sheltering the little house which had been our home for four months. Bathing maidens waved their towels in farewell.

Rangitoto Beacon sped by. Three hours later we sailed through the tide rip in Whangaparoa passage, and as dusk settled upon the surroundings it began to blow from eastward. Teddy rolled heavily.

The cabin was flooded with farewell presents; in the galley paper bags and packages of stores tumbled about. They should have been emptied into boxes and tins and properly stowed before we went to sea. In the bustle of farewells we had found no time for such details. Truly, we were hardly ready for sea. Also we were tired. We decided to run into Mansion House Bay at Kawau for the night.

Kawau is a beautiful little island some thirty miles from Auckland and a favourite goal for week-end excursions. There is a little pier in Mansion House Bay, alongside of which I could moor the boat, without going to the trouble of anchoring. This latter was an essential consideration, seeing that the anchor was securely lashed in the forepeak and the chain stowed away in the stern.

When, about ten at night, we made Teddy fast to the pier, it was pitchy dark. Kawau had gone to sleep.

Rising the next morning we found that the easterly had developed into a hard gale from the north-east. We were in no particular hurry to get to Brisbane, our next destination, so, instead of roughing it outside, we remained in the sheltered comfort of Mansion House Bay.

For three days the gale lasted, but by Wednesday morning, the 9th of March, it had exhausted itself and we departed. Leaving the shelter of the cove, we found a disappointingly light breeze of southerly wind blowing outside. The tides were at their maximum, running strongly northward and setting us backwards almost as fast as we could beat to windward against the light wind.

However, after several hours of sailing, we contrived to weather the reefs and thence lay south of Kawau, heading well to windward of the southern point of Challenger Island, a little rocky islet a quarter of a mile long in a north and south direction, which is separated from the south-east end of Kawau by a narrow channel.

The gale had left a heavy swell, which broke thunderously over the rocky ledges of the point.

The breeze seemed to freshen a little as we were approaching these rocks, but the tide was setting strongly to leeward, so that when we were within a hundred yards off the point, it was obvious that we could not weather it. The point then lay east of us, and Teddy headed south-east. I therefore put the helm down in order to go about.

Strange! She would not obey the helm.

The wind had suddenly died out.

However, Teddy was still moving ahead; she surely had sufficient headway to respond to the rudder, and there was no sea. Here, on the western side of the point the water was almost smooth. Smooth indeed, abominably smooth! Glassy, like the polished surface of a river, where it hastens towards a precipice.

I tried again—drove the tiller hard to leeward, again and again. No response! Consternation seized me: the current had Teddy in her power!

With accelerating speed we were driven towards the point, on the other side of which the swell rose into gigantic breakers, which, hurling themselves against the rugged obstacles with thundering fury sent rumbling waterfalls of foam over the rocky ledges. Sunken rocks off the point showed their frothy fangs, thirty, twenty yards away. The tumult was deafening. Oh, how I hated them, those rocks, those breakers, those snarling fangs, threatening, sneering, evil, inevitable...

I rushed forward and ran out the heavy sweep, tore and pulled and rowed with impotent rage. My wife cast off the halyards. The sails came clattering down.

Like a mill-race the current swept round the point.

Then the sweep broke.

I grabbed the spinnaker-boom in a foolish attempt to stop a weight of twenty five tons driven onward at five knots speed. A desperate man will do stupid things.

Now, we were close against it. We felt the lift of the surge: cold breaths of a moisture-laden atmosphere chilled us. My heart shrunk within me: *Teddy*'s end was near.

We struck the first time. I felt how the rocks crunched beneath our keel. Teddy heeled over, hard, then, righting herself, was lifted again and carried onward, past the point, right into the breakers...

I shouted to my wife to fetch little Tui from out of her bunk in the cabin. The same instant Teddy was seized by an enormous wave, lifted high, and with one big sweep thrown sideways against the rugged rocks. Then everything seemed to happen at the same time. Planks crushed, spars splintered. Rumbling—crashing—shrieking—rushing waters—and above it all the thundering roar of ten thousand unfettered demons of the waves.

Sometimes buried in foam so that we lost our breath, sometimes clinging to an almost perpendicular deck, when the sea was on the return, we needed all the presence of mind that we could muster. The main boom had come adrift. One of the topping lifts had jammed in a block sheave aloft, keeping the heavy spar suspended just over the cabin coamings and leaving it to sweep wildly from side to side, hardly twenty inches above the deck, whenever the boat rolled. The boom was like a huge club swung with deadly intent by a cunning giant hand. To dodge its shattering blows, we were continually forced to flatten out on the foam-swept deck.

Tony in his canvas harness, tied by a short rope to the rail, was in immediate danger. I succeeded in undoing his rope and then, in that fraction of a second when the main boom hung still, while gathering momentum for another sweeping assault, I ventured to leap onto the rugged face of the rock. It was a desperate chance. For one terribly long moment I hung by one hand, with the other attempting to support Tony, whose grip around my neck was gradually loosening, while the rush of returning water seemed to load insufferable tons of weight on to us. The next moment I found a foothold and climbed onto the ledge, where I left Tony with strict orders to grab a hold and hang on. Then I returned to the boat.

Again and again the receding surf would drag the doomed boat away from the cliff, brutally tearing her trembling timbers over an uneven rocky bottom, bumping her, ripping her with jagged teeth and leaving her heeling to seaward with the lower part of a precipitous deck submerged in surging foam and the mainboom-end swinging insanely about in the milky whirlpool. Then the next wave would pick her up and dash her against the rocks with the full force of her own weight. Oh, how I suffered!

I had taken Tui from Julie and told the latter to jump ashore, as soon as a chance offered. With my wife ashore to receive the baby, our chances of saving her would be all the better, I judged.

My wife slipped, or jumped short, or was washed overboard, none of us knows exactly how it happened. I only knew that I saw Julie disappear in the seething foam between the boat and the rocks. I saw my wife being whirled about in the churning, roaring surf amid countless jagged spikes protruding from the rocks. At intervals I saw her head, a foot or an arm above the seething waters, now close to the cliff, now away out. I saw her clinging to the rocks in a desperate attempt to climb up even while the next breaker came thundering in. I pointed seaward, shouting for her to try to swim clear of the surf. Then the breaker was upon her. I could do nothing until I had saved Tui. There was no place on the boat where I could leave the baby even for ten seconds without committing her to certain death. As it was, I had sufficient reason to fear that Tui would drown in my arms ere I could bring her ashore. But it is hard to see one's best friend fighting a desperate battle for life without being able to come to her assistance.

However, Julie had not been battered to death against the rocks. She came to the surface again and grabbed hold of the mainsheet, but before I could tend her a helping hand, the boom ran out jerking tight the sheet and hurling my wife away.

Somehow I contrived to bring Tui ashore. Hurriedly I handed her to Tony, instructing him to look after her and not to leave her, no matter what happened. I then hastened to the rescue of my Julie.

But in the meantime she had achieved the seemingly impossible feat of swimming clear of the surging breakers. With infinite relief I saw her making for a large piece of floating timber on the sheltered side of the point, meantime using the broken sweep for a support. A motor launch I had not previously noticed, had put out a dinghy to go to her assistance. She called out to me in a voice which showed that she was very much alive. Well, she had certainly proved her mettle.

Tony, too, had behaved like a hero. Without a word he had taken his duties like a man, sitting on the rocks where I had left him in charge of Tui. He never budged, even when the breakers washed over them occasionally. In all my misery I could not help feeling proud of him.

Spare Provisions had also been washed overboard. I had seen the poor dog fighting in the surf, but could do nothing for her. However, she had saved herself; limping and bleeding she joined our little group of castaways.

The fishing launch brought us back to Kawau.

When the family had been put up at the Mansion House, I returned to the wreck to see if anything could be salvaged. It proved impossible; the seas were continually sweeping over the hull, which was waterlogged and badly strained. Her flag, the proud ensign of the Royal Norwegian Yacht Club, awarded her for her merits, was still flying, half mast, where it had jammed in the rush of events. 'As if *Teddy* bemoaned her own end!' said one of the fishermen. I felt differently. To me it seemed as if my noble boat mourned not her own funeral but the end of a beautiful dream and the misfortune of the master, who loved her.

I turned away. I had seen her dear and pretty lines for the last time.

When I returned on the following morning, *Teddy*—my kingdom—had vanished.

EPILOGUE

Never has a finer craft existed.

Graceful were her lines, ever pleasing to the eye, because she was the embodiment of usefulness. Like a true masterpiece she stood above the fickle taste of fashion.

She was fast and she was safe. She was an able boat. During the thirty-five years when she served as a pilot boat, surely many a ship's crew and many a valuable cargo would have been lost but for the staunch ability of this boat.

Now she has gone. I have cried over her like a child, wept over her as over the loss of a dear friend. Heartsick I have stumbled about amongst the wreckage washed ashore on Challenger Island the day after the disaster, picking up odd bits of timber here and there and pressing them to my heart. Each piece I recognized, each spoke to me, each told me a tale.

In a hidden cove a square piece of planking had driven ashore. I sat down on it to rest, caressing those planks with sorrowing hands, and then, glancing warily around lest someone should see me, I bent down and kissed those planks good-bye.

Oh, I am not ashamed of that now.

She has gone, our boat, our home for nigh on four years. Splendid as had been her career as a pilot boat, she ended her days as the plaything of a foolish seeker of happiness, a fool, however, with a great love in his heart for the boat that carried him as near as it is given mortals to come to ever-elusive happiness. Can you wonder that I loved her?

I have told you the tale of our cruise, and I find, now, that the tale of our cruise is in reality the tale of a noble boat, who, like a faithful dog or like an ageing horse, showed loyalty and love even to the master who misused her.

Dear old *Teddy*!